999

Nonquantitative Problems

for FE Examination Review

Kenton Whitehead, PhD, PE

Professional Publications, Inc. • Belmont, California

Production Manager: Aline M. Sullivan
Acquisitions Editor: Gerald R. Galbo
Project Editor: Mia Laurence
Copy Editor: Mia Laurence
Book Designer: Yvonne M. Sartain
Typesetter: Cathy Schrott
Illustrator: Yvonne M. Sartain
Proofreader: Margaret S. Yoon
Cover Designer: Yvonne M. Sartain

999 NONQUANTITATIVE PROBLEMS FOR FE EXAMINATION REVIEW

Printed in the United States of America

Professional Publications, Inc.
1250 Fifth Avenue, Belmont, CA 94002
(415) 593-9119
www.ppi2pass.com

Current printing of this edition: 1

Library of Congress Cataloging-in-Publication Data
Whitehead, Kenton, 1939–
    999 Nonquantitative problems for FE examination review / Kenton
Whitehead.
        p.   cm.
    ISBN 1-888577-11-8 (perfect bound)
    1. Engineering--Examinations--Study guides. 2. Engineering--
Problems, exercises, etc.   I. Title.
TA159.W48   1997
620′.0076--dc21   97-12586
                                                    CIP

Table of Contents

Professional Publications, Inc. • Belmont, California

Preface and Acknowledgments

999 Nonquantitative Problems for FE Examination Review was written as a companion volume to Michael Lindeburg's *1001 Solved Engineering Fundamentals Problems*. This volume covers the types of problems on the FE exam that do not require numerical calculations. It requires a qualitative understanding of the theories, principles, and definitions the successful examinee must know.

I selected the content for this publication based on my experiences teaching the Engineer-In-Training review class for the Professional Engineering Institute, working problems for these classes, and reviewing publications of the National Council of Examiners for Engineering and Surveying. All questions are relevant to the current exam format.

The problems are divided similarly to those on the exam. The number of questions on each subject are in rough proportion to the number of questions on the exam. All material used in the questions and answers stem from Michael Lindeburg's *Engineer-In-Training Reference Manual*.

I chose all of the questions myself so that the relevance, degree of ease (or difficulty), and breadth of scope of questions written by a single author would be consistent. I hope I have achieved that goal.

The beginning of this project can be traced to an idea of Michael Lindeburg and its implementation by acquisitions editor Mary Fiala.

I wish to thank my wife Carole who supported my long evenings and weekends of work on this effort. Much useful criticism was offered by our daughters Hilary and Amy, who, as college students, recognized and helped me appreciate unambiguous questions.

All the answers have been reviewed for content, accuracy, and relevance by technical experts, including Steve Van Wyk, PE, Instructor at Olympic College. I thank them. I also offer my thanks to the acquisitions and production staffs of Professional Publications.

The work was a pleasant adventure, and I hope it will prove a useful addition to the fine books published by Professional Publications. I hope you will find it a valuable aid to your review.

Kenton Whitehead, PhD, PE
Poway, CA

How to Use this Book

If you never read the material at the front of your books anyway, and if you are in a hurry to begin and you only want to read one paragraph, here it is:

> All chapters in this book are independent. Start with any one. Answer all the questions that you have time for. Don't peek at the answers until you've thought about the question for a while. Answer questions from your weak areas as well as your strong areas. Keep studying until the exam. Good luck!

However, if you want a thorough review, you will probably want to know a little more about reviewing for the exam. The rest of this page is for you.

In the "good old days," when examinations for engineering licensing were in their infancy, most review books were mainly compilations of problems with little supporting theory—much like this one. Such books placed the burden on the examinees to accumulate and become familiar with numerous textbooks and references.

The National Council of Examiners for Engineering and Surveying (NCEES) changed all that when it limited you to a single reference booklet which it provides. While this eliminated the "shopping cart syndrome" (wherein some examinees literally brought shopping carts of books to the exam), it also changed the nature of how most examinees review. Instead of having to accumulate numerous books, the burdens on the examinee are now (a) memorizing the engineering concepts that they can no longer look up in their textbooks, (b) learning to use the NCEES-supplied reference document, and (c) learning to work problems quickly. Working countless practice problems has become the review method of choice for many examinees.

The scope of the exam has also been narrowed significantly, and examinees no longer review a wide variety of engineering subjects. The scope of their review has narrowed accordingly, placing more emphasis on the ability to work certain standard types of problems quickly. This book contains the grist for the standard problem review mill.

This book doesn't contain any supporting theory and is not meant to be used as a stand-alone exam review. Most of the answers to the questions are short and sweet. That is because this book was meant to be used in conjunction with two other references: a FE exam review book (for example, either the *EIT Review Manual* or *Engineer-In-Training Reference Manual*) and the *NCEES Reference Handbook* (all are available through Professional Publications). You should keep these by your side while you are studying.

Depending on your preference, you might decide to first review a subject and then test your comprehension with the questions in this book. Or you might decide to jump right in and try to answer the questions, reviewing only those subjects that you are rusty in or unfamiliar with. This book can be used either way, although the former way really is better.

While working through the questions, you should try to use the *NCEES Reference Handbook* as your sole reference source. Refer to the *EIT Review Manual* or *Engineer-In-Training Reference Manual* when you need to refresh your memory about subjects that have become dim. However, use the *NCEES Reference Handbook* when you simply need a formula or data. That way you will become intimately familiar with the only reference that is permitted in the examination room.

The National Society of Professional Engineers

Whether you design water works, consumer goods, or aerospace vehicles; whether you work in private industry, for the U.S. government, or for the public; and whether your efforts are theoretical or practical, you (as an engineer) have a significant responsibility.

Engineers of all types perform exciting and rewarding work, often stretching new technologies to their limits. But those limits are often incomprehensible to nonengineers. As the ambient level of technical sophistication increases, the public has come to depend increasingly and unhesitatingly more on engineers. That is where professional licensing and the National Society of Professional Engineers (NSPE) become important.

NSPE, the leading organization for licensed engineering professionals, is dedicated to serving the engineering profession by supporting activities, such as continuing educational programs for its members, lobbying and legislative efforts on local and national levels, and the promotion of guidelines for ethical service. From local, community-based projects to encourage top-scoring high school students to choose engineering as a career, to hard hitting lobbying efforts in the nation's capital to satisfy the needs of all engineers, NSPE is committed to you and your profession.

Engineering licensing is a two-way street: it benefits you while it benefits the public and the profession. For you, licensing offers a variety of benefits, ranging from peer recognition to greater advancement and career opportunities. For the profession, licensing establishes a common credential by which all engineers can be compared. For the public, a professional engineering license is an assurance of a recognizable standard of competence.

NSPE has always been a strong advocate of engineering licensing and a supporter of the profession. Professional Publications hopes you will consider membership in NSPE as the next logical step in your career advancement. For more information regarding membership, write to the National Society of Professional Engineers, Information Center, 1420 King Street, Alexandria, VA 22314, or call (703) 684-2800.

Professional Publications, Inc. • Belmont, California

Chapter 1
Engineering Units and Mathematics

Problem–1

Each of the following is a common unit of mass except the

    (a) gram.
    (b) kilogram.
    (c) poundal.
    (d) slug.

The answer is (c)

Problem–2

In the English Engineering System, the units for force, mass, and acceleration used in Newton's law, $F = ma$, are given respectively by

    (a) lbm, slug, and m/s^2.
    (b) dyne, g, and cm/s^2.
    (c) lbf, lbm, and ft/sec^2.
    (d) N, kg, and m/s^2.

The answer is (c)

Problem–3

In the English Gravitational System, the units for force, mass, and acceleration used in Newton's law, $F = ma$, are given respectively by

    (a) lbf, slug, and ft/sec^2.
    (b) dyne, g, and cm/s^2.
    (c) lbf, lbm, and ft/sec^2.
    (d) N, kg, and m/s^2.

The answer is (a)

Problem–4

In the cgs system, the units for force, mass, and acceleration used in Newton's law, $F = ma$, are given respectively by

    (a) lbm, slug, and m/s^2.
    (b) dyne, g, and cm/s^2.
    (c) lbf, lbm, and ft/sec^2.
    (d) N, kg, and m/s^2.

The answer is (b)

Problem–5

Which of the following represents the greatest mass?

    (a) 454 g
    (b) 1 lbm
    (c) 1 kg
    (d) 1 slug

The answer is (d)

Problem–6

In SI (mks) units, the units for force, mass, and acceleration used in Newton's law, $F = ma$, are given respectively by

    (a) lbm, slug, and m/s^2.
    (b) dyne, g, and cm/s^2.
    (c) lbf, lbm, and ft/sec^2.
    (d) N, kg, and m/s^2.

The answer is (d)

Problem–7

All of the following are generally accepted types of units except

   (a) base units (meter, second, ampere).
   (b) derived units (N/m^2, J/s, W/m^2).
   (c) supplementary units (radians, steradians).
   (d) derived units with unusual names (photons/fortnight, skosh, pinch).

The answer is (d)

Problem–8

In any system of units, all of the following are base units except

   (a) length.
   (b) mass.
   (c) time.
   (d) steradian (solid angle).

The answer is (d)

Problem–9

A system of units used in engineering calculations is said to be consistent if

   (a) it is approved by the National Society of Professional Engineers.
   (b) no conversion units are required.
   (c) it is used throughout the Fundamentals of Engineering exam.
   (d) it is either the English Engineering or the SI system.

The answer is (b)

Problem–10

The system of units that is consistent is the

   (a) English Engineering system.
   (b) SI system.
   (c) Absolute English system.
   (d) cgs system.

The answer is (b)

Problem–11

A means of obtaining an equation describing a physical phenomenon without actually understanding the physical mechanism is

   (a) dimensional analysis.
   (b) Laplace transforms.
   (c) algebra.
   (d) tensor calculus.

The answer is (a)

Problem–12

The equation expressing the relationship between groups of dimensionless variables G_i, for example, $f(G_1, G_2, \ldots G_k) = 0$, is given by

   (a) the Buckingham π theorem.
   (b) the first law of thermodynamics.
   (c) Newton's second law.
   (d) Fourier's law.

The answer is (a)

Problem–13

Weight is the force exerted on an object due to its placement in a(n)

   (a) magnetic field.
   (b) gravitational field.
   (c) vacuum.
   (d) electric field.

The answer is (b)

Problem–14

The prefix *tera* denotes a multiple of

   (a) 10^{-12}.
   (b) 10^{-6}.
   (c) 10^6.
   (d) 10^{12}.

The answer is (d)

Problem–15

The prefix *nano* denotes a multiple of

  (a) 10^{-6}.
  (b) 10^{-9}.
  (c) 10^{-12}.
  (d) 10^{-15}.

The answer is (b)

Problem–16

The number of significant digits in the number 3.4050 is

  (a) 2.
  (b) 3.
  (c) 4.
  (d) 5.

The answer is (d)

Problem–17

The number 3.840×10^5 has how many significant digits?

  (a) 2
  (b) 3
  (c) 4
  (d) 5

The answer is (c)

Problem–18

The quotient of 27,372 and 3.84 is most correctly written as

  (a) 7.13×10^3.
  (b) 7128.
  (c) 7.128×10^3.
  (d) 7128.1.

The answer is (a)

Problem–19

The sum of the numbers 0.2056, 2.572, 14.25, and 576.1 is most correctly written as

  (a) 593.
  (b) 593.1.
  (c) 593.13.
  (d) 593.128.

The answer is (b)

Problem–20

Nonterminating, nonrepeating numbers that cannot be expressed as the ratio of two integers are

  (a) real numbers.
  (b) complex numbers.
  (c) rational real numbers.
  (d) irrational real numbers.

The answer is (d)

Problem–21

Imaginary numbers consist of

  (a) numbers that can be written as the ratio of two integers.
  (b) numbers that cannot be written as the ratio of two integers.
  (c) square roots of negative numbers.
  (d) nonterminating, nonrepeating numbers that cannot be expressed as the ratio of two integers.

The answer is (c)

Problem–22

Complex numbers consist of

  (a) numbers that can be written as the ratio of two integers.
  (b) numbers that cannot be written as the ratio of two integers.
  (c) square roots of negative numbers.
  (d) combinations of real and imaginary numbers.

The answer is (d)

Problem–23

The coefficients of a polynomial generated by the binomial theorem can be found from

    (a) the method of partial fractions.
    (b) the quadratic equation.
    (c) Pascal's triangle.
    (d) Descartes' rule of signs.

The answer is (c)

Problem–24

The maximum number of positive (or negative) real roots that a polynomial will have as determined by counting the number of its sign reversals is given by

    (a) the method of partial fractions.
    (b) the quadratic equation.
    (c) Pascal's triangle.
    (d) Descartes' rule of signs.

The answer is (d)

Problem–25

The method used to simplify a proper polynomial fraction of two polynomials whose numerator is one power less than the denominator is

    (a) the method of partial fractions.
    (b) the quadratic equation.
    (c) Pascal's triangle.
    (d) Descartes' rule of signs.

The answer is (a)

Problem–26

The number π is described most correctly as a

    (a) real number.
    (b) imaginary number.
    (c) rational real number.
    (d) irrational real number.

The answer is (d)

Problem–27

All of the following are generally recognized methods for solving simultaneous linear equations except

    (a) graphing.
    (b) substitution.
    (c) method of undetermined coefficients.
    (d) Cramer's rule.

The answer is (c)

Problem–28

When a complex number is written in the form $Z = a + bi$, it is expressed in

    (a) rectangular form.
    (b) phasor form.
    (c) cis form.
    (d) exponential form.

The answer is (a)

Problem–29

When a complex number is written in the form $Z = re^{i\theta}$ where $r = a^2 + b^2 = \text{mod}(Z)$ and $\theta = \arg(Z) = \arctan(b/a)$, it is expressed in

    (a) rectangular form.
    (b) phasor form.
    (c) cis form.
    (d) exponential form.

The answer is (d)

Problem–30

When a complex number is written in the form $Z = r\angle\theta$, it is expressed in

    (a) rectangular form.
    (b) trigonometric form.
    (c) phasor form.
    (d) exponential form.

The answer is (c)

Problem—31

When a complex number is written in the form $Z = r(\cos\theta + i\sin\theta)$, it is expressed in

- (a) rectangular form.
- (b) trigonometric form.
- (c) phasor form.
- (d) exponential form.

The answer is (b)

Problem—32

The expression relating the equality of complex numbers expressed in exponential and trigonometric form, $e^{i\theta} = \cos\theta + i\sin\theta$, is called

- (a) Pascal's triangle.
- (b) Euler's equation.
- (c) Bernoulli's equation.
- (d) the quadratic equation.

The answer is (b)

Problem—33

One good practical reason for using multiple forms of complex numbers is

- (a) phasor forms are used in the English Engineering System and rectangular forms are used in the SI.
- (b) rectangular forms are used in the English Engineering System and phasor forms are used in the SI.
- (c) polar and phasor forms are easier for division and multiplication of complex numbers, and rectangular forms are easier for addition and subtraction.
- (d) rectangular forms are easier for division and multiplication of complex numbers, and polar and phasor forms are easier for addition and subtraction.

The answer is (c)

Problem—34

Division of one complex number by another when both are expressed in rectangular form is typically performed by using

- (a) an engineering calculator.
- (b) the complex conjugate (rationalizing).
- (c) the method of undetermined coefficients.
- (d) Pascal's triangle.

The answer is (b)

Problem—35

The value a function approaches when the independent variable in the function approaches a target value is called

- (a) a limit (limiting value).
- (b) a finite difference.
- (c) the mean value.
- (d) the tangent.

The answer is (a)

Problem—36

L'Hôpital's rule should be used when

- (a) the numerator and denominator of an algebraic expression both approach zero or both approach infinity.
- (b) the numerator of an algebraic expression approaches zero but the denominator approaches infinity.
- (c) the denominator of an algebraic expression approaches zero but the numerator approaches infinity.
- (d) a limit can not be found any other way.

The answer is (a)

Problem-37

A sequence is

    (a) the sum of terms in a series.

    (b) an ordered progression of numbers.

    (c) a method of determining convergence of an un-known series.

    (d) any gathering of a finite number of terms.

The answer is (b)

Problem-38

Each of the following names describes a standard sequence except

    (a) geometric sequence.

    (b) arithmetic sequence.

    (c) q-sequence.

    (d) p-sequence.

The answer is (c)

Problem-39

A series is the sum S_n of n terms in a

    (a) sequence.

    (b) arithmetic expression.

    (c) polynomial.

    (d) improper fraction.

The answer is (a)

Problem-40

A series is said to converge (be convergent) if the sum S_n

    (a) is finite.

    (b) exists, as $n \to \infty$.

    (c) is infinite.

    (d) is less than one.

The answer is (b)

Problem-41

A finite series has

    (a) a finite number of terms.

    (b) fewer than 100 terms.

    (c) successive terms of alternating signs.

    (d) successive terms with the same sign.

The answer is (a)

Problem-42

What is the name given to the following test to determine whether a series converges or diverges?

$$L = \lim_{n \to \infty} \frac{a_{n+1}}{a_n} \begin{cases} < 1 \text{ converges} \\ = 1 \text{ inconclusive} \\ > 1 \text{ diverges} \end{cases}$$

    (a) the comparison test

    (b) the ratio test

    (c) L'Hôpital's rule

    (d) the mean value theorem

The answer is (b)

Problem-43

Comparing successive terms of an unknown series to those of a series that is known to converge or diverge is called

    (a) the comparison test.

    (b) the ratio test.

    (c) L'Hôpital's rule.

    (d) the mean value theorem.

The answer is (a)

Problem–44

If a series containing alternating signs is equal term for term to a converging series with all positive signs, the series is said to be

    (a) monotonically increasing.
    (b) absolutely convergent.
    (c) conditionally convergent.
    (d) finite.

The answer is (b)

Problem–45

If an all-positive term diverging series is equal term for term to a converging series with alternating signs, the original series is said to be

    (a) monotonically increasing.
    (b) absolutely convergent.
    (c) conditionally convergent.
    (d) finite.

The answer is (c)

Problem–46

Which of the following is not a correct derivative?

    (a) $\dfrac{d}{dx}e^{-2x} = -2e^{-2x}$

    (b) $\dfrac{d}{dx}(\ln 2x) = \dfrac{1}{2x}$

    (c) $\dfrac{d}{dx}(\sin 2x) = 2\cos 2x$

    (d) $\dfrac{d}{dx}(1-x)^2 = (-2)(1-x)$

The answer is (b)

Problem–47

What is the maximum value of the following function?

$$y = -2x^2 + 4x + 3$$

    (a) 2
    (b) 3
    (c) 4
    (d) 5

The answer is (d)

Problem–48

What is the following integral?

$$\int \frac{3x^3 + 2x^2 + 4}{x}\,dx$$

    (a) $6x + 2 + 4x^{-2} + C$
    (b) $x^2 + 2x + \ln 4x + C$
    (c) $x^3 + x^2 + \ln x^4 + C$
    (d) $\ln(3x^3 + 2x^2 + 4) + C$

The answer is (c)

Problem–49

Evaluate the following limit as $x \to 0$.

$$\lim_{x \to 0} \frac{1 - \cos^2 x}{\sin x}$$

    (a) 0
    (b) $\frac{1}{2}$
    (c) 1
    (d) 2

The answer is (a)

$$8+8 - \frac{8}{3} - \left(-\frac{8}{3}\right) = 16 \div \frac{16}{3} = 16 \cdot 5\frac{1}{3}$$

Problem–50

$$\int_{-2}^{+2} 4-x^2 \, dx = 4x\Big|_{-2}^{+2} - \frac{x^3}{3}\Big|_{-2}^{+2}$$

What is the area enclosed by the function $y = 4 - x^2$ and the x-axis, between $x = -2$ and $x = +2$?

(a) 4
(b) 8
(c) $10\frac{2}{3}$
(d) 15

The answer is (c)

Problem–51

What type of differential equation is shown?

$$y = e^{-8x} + yy' - 2y''.$$

(a) nonlinear, second-order, and nonhomogeneous
(b) linear, second-order, and nonhomogeneous
(c) linear, second-order, and homogeneous
(d) nonlinear, first order, and homogeneous

The answer is (a)

Problem–52

Given that A and B are constants, which answer is a solution of the following differential equation?

$$y'' + 10y' + 9y = 0$$

(a) $y = Ae^{10x} + Be^{-x}$
(b) $y = Ae^{-4x} + Be^{-5x}$
(c) $y = Ae^{9x} + Be^x$
(d) $y = Ae^{-9x} + Be^{-x}$

The answer is (d)

$$(y+1)(y+9) = 0$$
$$y = -1$$
$$y = -9$$

Problem–53

What is the general solution to the following differential equation?

$$\frac{d^2y}{dx^2} + 2\frac{dy}{dx} + 2y = 0$$

$(y+1)^2$
$y = -1$

(a) $y = A\sin x + B\cos x$
(b) $y = A\sin 2x + B\cos 2x$
(c) $y = A\sin \sqrt{2}x + B\cos \sqrt{2}x$
(d) $y = e^{-x}(A\sin x + B\cos x)$

The answer is (d)

Problem–54

The intersection of two straight lines $y_1 = 8x + 5$ and $y_2 = 12 - 6x$ occurs when x equals

(a) $\frac{1}{2}$.
(b) 1.
(c) 2.
(d) 4.

$8x+5 = 12-6x$
$14x = 12-5 = 7$
$x = \frac{1}{2}$

The answer is (a)

Problem–55

What is the equation of the line that goes through points $(4,-6)$ and $(0,-5)$?

(a) $y = mx - 20$
(b) $y = -\frac{1}{4}x - 5$
(c) $y = \frac{1}{5}x + 5$
(d) $y = \frac{1}{4}x + 5$

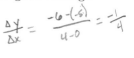

$\frac{\Delta y}{\Delta x} = \frac{-6-(-5)}{4-0} = -\frac{1}{4}$

The answer is (b)

Problem–56

What are the solutions to the following simultaneous linear equations?

$$10x + 3y + 10z = 26$$
$$8x - 2y + 9z = 11$$
$$8x + y - 10z = 36$$

(a) $x = 1$, $y = 2$, $z = 3$
(b) $x = -1$, $y = 3$, $z = 2.7$
(c) $x = 3.2$, $y = 2$, $z = -1.2$
(d) $x = 3$, $y = 2$, $z = -1$

The answer is (d)

Problem–57

In the expansion of the polynomial fraction into partial fractions shown below, the values of A and B are

$$\frac{3x + 2}{x^2 + 3x + 2} = \frac{A}{x + 2} + \frac{B}{x + 1}$$

$$A(x+1) + B(x+2) = 3x+2$$

(a) $A = 2$, $B = 1$
(b) $A = 3$, $B = 2$
(c) $A = 4$, $B = -1$
(d) $A = 1$, $B = 3$

The answer is (c)

Problem–58

What is the partial fraction expansion of $\mathcal{L}(s)$?

$$\mathcal{L}(s) = \frac{30}{s(s + 3)}$$

(a) $\dfrac{30}{s} - \dfrac{1}{s + 3}$

(b) $\dfrac{10}{s} - \dfrac{10}{s + 3}$

(c) $\dfrac{3}{s} - \dfrac{10}{s + 3}$

(d) $\dfrac{10}{s} + \dfrac{10}{s + 3}$

The answer is (b)

Problem–59

What is the usefulness of the Student's t-test?

(a) testing the distribution of outcomes to see if they come from a normal distribution
(b) comparing values about the mean
(c) determining if the probability function is symmetric about zero
(d) determining if the difference between two samples is significant

The answer is (d)

Problem–60

The area under the standard normal curve between the mean and one standard deviation ($z = 1$) is

(a) 34% on either side of the mean.
(b) 25% of the area under the curve.
(c) 17% on either side of the mean.
(d) The standard deviation is not used.

The answer is (a)

Problem–61

What is the probability of a fair coin showing at least one head in seven flips?

(a) $\left(\frac{1}{2}\right)^7$

(b) $\frac{1}{2} + \left(\frac{1}{2}\right)^7$

(c) $\dfrac{7!}{1!6!}\left(\frac{1}{2}\right)^7$

(d) $1 - \left(\frac{1}{2}\right)^7$

The answer is (d)

Problem–62

The cross-product $\vec{A} \times \vec{B}$ of $\vec{A} = 3\mathbf{i} + 4\mathbf{j}$ and $\vec{B} = 4\mathbf{i} - 3\mathbf{j}$ is

(a) $7\mathbf{k}$.
(b) $-25\mathbf{k}$.
(c) $\mathbf{i} - 7\mathbf{j} + 4\mathbf{k}$.
(d) $12\mathbf{i} - 12\mathbf{j}$.

The answer is (b)

Problem–63

What is the result of dividing $7 + 5j$ by $3 + 4j$ ($j = \sqrt{-1}$)?

(a) $\dfrac{41 - 13j}{25}$

(b) $2 - \frac{1}{2}j$

(c) $\dfrac{21}{25} + \dfrac{13}{25}j$

(d) $\dfrac{28}{25}$

The answer is (a)

Problem–64

The equation $x^2 + y^2 + 12y - 2x = -12$ describes

(a) a hyperbola centered at $(1, -6)$.
(b) an ellipse of semi-major axis $\sqrt{12}$.
(c) a circle of radius 5.
(d) a parabola of directrix 3.

The answer is (c)

Problem–65

What is the distance between points $(3, 2, -1)$ and $(4, 5, 0)$?

(a) 5
(b) 9
(c) $\sqrt{17}$
(d) $\sqrt{11}$

The answer is (d)

Problem–66

The matrix product $\mathbf{AB}$ of $\mathbf{A} = \begin{bmatrix} 2 & 1 \\ 1 & 0 \end{bmatrix}$ and $\mathbf{B} = \begin{bmatrix} 4 & 3 \\ 2 & 1 \end{bmatrix}$ is

(a) $\begin{bmatrix} 12 & 9 \\ 2 & 1 \end{bmatrix}$

(b) $\begin{bmatrix} 10 & 7 \\ 4 & 3 \end{bmatrix}$

(c) $\begin{bmatrix} 10 & 4 \\ 3 & 9 \end{bmatrix}$

(d) 2

The answer is (b)

Problem–67

What is the *volume* of the rectangular solid defined by the zero-based vectors $\mathbf{A}$, $\mathbf{B}$, and $\mathbf{C}$?

$$\mathbf{A} = 2\mathbf{i} - 2\mathbf{j} + \mathbf{k}$$
$$\mathbf{B} = 4\mathbf{i} + 2\mathbf{j} + 2\mathbf{k}$$
$$\mathbf{C} = \mathbf{i} + 5\mathbf{j} + 4\mathbf{k}$$

(a) 14
(b) 28
(c) 56
(d) 82

The answer is (c)

Problem–68

All of the following are excellent reasons to employ simulation for modeling a system except

(a) safety.
(b) expense.
(c) unavailability of appropriate analytical techniques.
(d) technical naiveté.

The answer is (d)

Problem–69

Which of the following is not a desirable property of a random number generator?

  (a) The numbers should come from a uniform distribution.
  (b) The generation should be fast.
  (c) The sequence should not repeat.
  (d) The sequence should have a varying number of significant figures.

The answer is (d)

Problem–70

All of the following are recognized and accepted methods of general systems modeling except

  (a) Monte Carlo simulation.
  (b) replacement and renewal models.
  (c) decision trees.
  (d) rule of thumb methods.

The answer is (d)

Problem–71

All of the following units can be converted to an equivalent energy of British thermal units (Btu) except

  (a) calories.
  (b) carats.
  (c) foot-pounds.
  (d) joules.

The answer is (b)

Problem–72

All of the following units can be converted to an equivalent charge except

  (a) Ampere-hours.
  (b) electrostatic units.
  (c) faradays.
  (d) farthings.

The answer is (d)

Problem–73

All of the following units can be converted to an equivalent angle except

  (a) hours.
  (b) minutes.
  (c) radians.
  (d) revolutions.

The answer is (a)

Problem–74

All of the following units can be converted to an equivalent power in dyne-cm/s except

  (a) Btu/min.
  (b) ft-lbf/sec.
  (c) J.
  (d) W.

The answer is (d)

Problem–75

All of the following units can be converted to an equivalent pressure in feet of water except

  (a) atmospheres.
  (b) bars.
  (c) kilopascals.
  (d) pounds.

The answer is (d)

Problem–76

All of the following units can be converted to an equivalent power in ft-lbf/min except

  (a) Btu/sec.
  (b) ergs/sec.
  (c) hp.
  (d) kg/s.

The answer is (d)

Chapter 2

Statics

Problem–1

A rigid body in static equilibrium experiences

    (a) only small forces.
    (b) only large forces.
    (c) no balanced forces.
    (d) no unbalanced forces.

The answer is (d)

Problem–2

All of the following are sources of external forces on rigid bodies except

    (a) physical contact (pushing).
    (b) gravity.
    (c) magnetism.
    (d) compressive and tensile forces holding the body together.

The answer is (d)

Problem–3

Internal forces on rigid bodies are due to

    (a) physical contact (pushing).
    (b) gravity.
    (c) electrostatics.
    (d) compressive and tensile forces holding the body together.

The answer is (d)

Problem–4

All of the following attributes characterize a force except

    (a) magnitude.
    (b) direction.
    (c) line of action.
    (d) center of rotation.

The answer is (d)

Problem–5

The moment due to an applied force on a body is zero only when

    (a) the force is negative.
    (b) the force is through the origin.
    (c) the line of action passes through the center of rotation.
    (d) the force is a function of time.

The answer is (c)

Problem–6

The moment of a force $\mathbf{F}$ applied at a distance $\mathbf{r}$ from a point O is equal to what quantity?

    (a) $\mathbf{M_O} = \mathbf{r} \cdot \mathbf{F}$
    (b) $\mathbf{M_O} = \nabla \cdot \mathbf{F}$
    (c) $\mathbf{M_O} = \mathbf{r} \times \mathbf{F}$
    (d) $\mathbf{M_O} = \nabla \times \mathbf{F}$

The answer is (c)

Problem–7

The line of action of the moment vector is _____ the plane of the force and position vectors.

    (a) within
    (b) parallel to
    (c) normal to
    (d) unrelated to

The answer is (c)

Problem–8

A couple is composed of a pair of forces that are

    (a) unequal, opposite, and nonparallel.
    (b) unequal, opposite, and parallel.
    (c) equal, opposite, and parallel.
    (d) equal and parallel forces.

The answer is (c)

Problem–9

If a force F is moved a distance d from the original point of application, to counteract the induced couple, one must add

    (a) nothing.

    (b) a force $F = M/d$.

    (c) a moment $M = Fd$.

    (d) a force couple located at a distance $d = F/M$ away.

The answer is (c)

Problem–10

A linear force system is shown in which part of the following illustration?

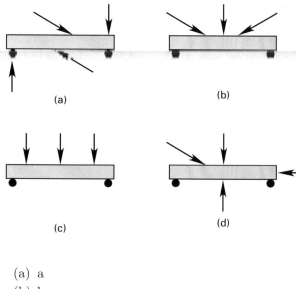

    (a) a

    (b) b

    (c) c

    (d) d

The answer is (c)

Problem–11

All the forces in a linear force system can be replaced by an equivalent resultant force of which of the following?

    (a) $F_R = \sum_i F_i$

    (b) $F_R = \dfrac{\sum_i F_i x_i}{x_R}$

    (c) $F_R = \dfrac{\sum_i F_i x_i}{\sum_i x_i}$

    (d) $F_R = \dfrac{M_R}{\sum_i F_i x_i}$

The answer is (a)

Problem–12

The equivalent force F_e equal to a continuously distributed load $w(x)$ is given by

    (a) $\displaystyle\int_0^L \dfrac{w(x)}{x}\,dx.$

    (b) $\displaystyle\int_{x=0}^{x=L} w(x)\,dx.$

    (c) $\displaystyle\int_{x=0}^{x=L} x\,w(x)\,dx.$

    (d) $\displaystyle\sum_i w_i(x).$

The answer is (b)

Problem–13

The point of application F_R where an equivalent force can be applied to give the same force and moment as a distributed load $w(x)$ is located at what position in the distributed load?

    (a) the middle

    (b) the centroid of the distributed load

    (c) the left end

    (d) the right end

The answer is (b)

Refer to the following illustrations for Probs. 14 through 18.

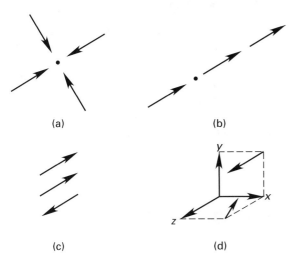

(a)

(b)

(c)

(d)

Problem–14

In the illustrations shown above, the only force system that is clearly not coplanar is

(a) a.
(b) b.
(c) c.
(d) d.

The answer is (d)

Problem–15

In the illustrations shown, a parallel force system is shown by

(a) a.
(b) b.
(c) c.
(d) d.

The answer is (c)

Problem–16

In the illustrations shown, a general three-dimensional force system is shown by

(a) a.
(b) b.
(c) c.
(d) d.

The answer is (d)

Problem–17

In the illustrations shown, a colinear force system is shown by

(a) a.
(b) b.
(c) c.
(d) d.

The answer is (b)

Problem–18

In the illustrations shown, a concurrent force system is shown by

(a) a.
(b) b.
(c) c.
(d) d.

The answer is (a)

Problem–19

A two-force member can be in static equilibrium if the two forces

(a) are colinear.
(b) are equal.
(c) are equal, opposite, and colinear.
(d) form a couple.

The answer is (c)

Refer to the following illustrations for Probs. 20 through 24.

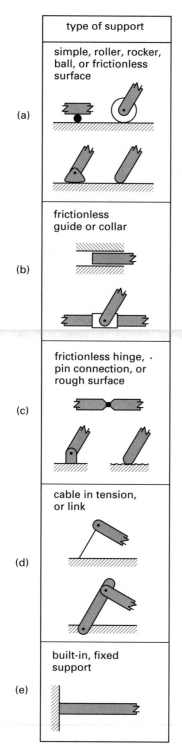

type of support
simple, roller, rocker, ball, or frictionless surface
(a)
frictionless guide or collar
(b)
frictionless hinge, · pin connection, or rough surface
(c)
cable in tension, or link
(d)
built-in, fixed support
(e)

Problem–20

The reaction(s) for the roller support in Fig. a are

(a) normal to the surface with no moment.
(b) in line with the cable, link, or member with no moment.
(c) normal to the rail, with no moment.
(d) two reaction components with one moment.

The answer is (a)

Problem–21

The reaction(s) for the frictionless guide or collar in Fig. b are

(a) normal to the surface with no moment.
(b) in line with the cable, link, or member with no moment.
(c) normal to the rail, with no moment.
(d) two reaction components with one moment.

The answer is (c)

Problem–22

The reaction(s) for the frictionless hinge, pin connection, or rough surface in Fig. c are

(a) normal to the surface with no moment.
(b) normal to the rail, with no moment.
(c) two reaction components with one moment.
(d) reactions in any direction with no moment.

The answer is (d)

Problem–23

The reaction(s) for the cable in tension or link in Fig. d are

(a) in line with the cable, link, or member with no moment.
(b) normal to the rail, with no moment.
(c) two reaction components with one moment.
(d) reactions in any direction with no moment.

The answer is (a)

Problem–24

The reaction(s) for the built-in, fixed support in Fig. e are

(a) normal to the surface with no moment.
(b) normal to the rail, with no moment.
(c) two reaction components with one moment.
(d) reactions in any direction with no moment.

The answer is (c)

Problem–25

Which of the following statements is not a characteristic of a statically indeterminant rigid body force system?

(a) It can be solved for all unknowns, which are usually reactions supporting the body.
(b) One or more of the members can be removed or reduced in restraint without affecting the equilibrium position.
(c) The number of redundant members is the degree of indeterminancy.
(d) A statically indeterminant body requires additional equations to supplement the equilibrium equations.

The answer is (a)

Problem–26

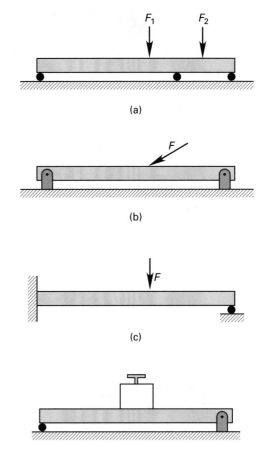

In the illustrations shown, all of the structures are statically indeterminant except which of the following?

(a) a
(b) b
(c) c
(d) d

The answer is (d)

Problem–27

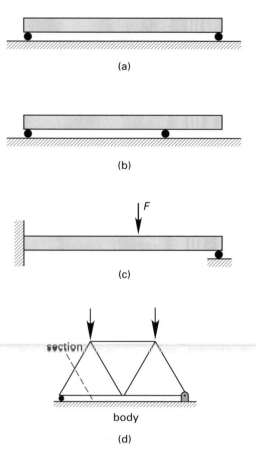

(a)

(b)

(c)

(d)

In the illustrations shown, all of the structures are statically determinant except which of the following?

    (a) a
    (b) b
    (c) c
    (d) d

The answer is (c)

Problem–28

A free moment is

    (a) a couple.
    (b) any moment without a specific point of application.
    (c) the time between questions on the FE exam.
    (d) a redundant moment removable from a statically indeterminant structure.

The answer is (a)

Problem–29

A hinge can support

    (a) a force but no moment.
    (b) a force and a moment.
    (c) neither a force nor a moment.
    (d) two forces and a moment.

The answer is (a)

Problem–30

The ratio of a load-bearing force to the applied force for a lever is called its

    (a) load factor.
    (b) efficiency.
    (c) mechanical advantage.
    (d) influence factor.

The answer is (c)

Problem–31

One primary advantage to applying a lever to a load is that it reduces the

    (a) load-bearing force.
    (b) applied force.
    (c) load-bearing energy.
    (d) distance through which the applied force moves.

The answer is (b)

Problem–32

A pulley is used to move a load for all the following reasons except to

    (a) reduce force.
    (b) reduce energy.
    (c) change direction of an applied tensile force.
    (d) gain a mechanical advantage.

The answer is (b)

Problem–33

Which of the following statements about axial members in a truss is not true?

  (a) They are capable of supporting axial forces only.
  (b) They are loaded only at their joints (ends).
  (c) They can support moments.
  (d) They disregard their own weight.

The answer is (c)

Problem–34

Which of the following statements about axial members is not true?

  (a) A horizontal member carries only horizontal loads. It cannot carry vertical loads.
  (b) A vertical member carries only vertical loads. It cannot carry horizontal loads.
  (c) The vertical components of an axial member's force is equal to the vertical component of the load applied to the member.
  (d) The end moments must sum to zero.

The answer is (d)

Problem–35

Which of the following formulas correctly describes the number of members (NOM) of a statically determinant truss?

  (a) number of members =
      (3) (number of joints) − 2
  (b) number of members =
      (2) (number of joints) − 3
  (c) number of members =
      (2) (number of joints) − 2
  (d) number of members =
      (2) (number of joints) − 1

The answer is (b)

Problem–36

All of the following are acceptable methods for determining loads in a statically determinant truss except

  (a) the cut-and-sum method.
  (b) the dummy unit load method.
  (c) the method of sections.
  (d) the superposition method.

The answer is (b)

Problem–37

Which of the following statements about the analysis of ideal cables carrying concentrated loads is not true?

  (a) The method of joints can be used.
  (b) The method of sections can be used.
  (c) All cables are in tension.
  (d) The horizontal loads in cable segments will be unequal.

The answer is (d)

Problem–38

A cable carrying a constant uniform load per unit length with respect to the horizontal axis will have the shape of a

  (a) straight line.
  (b) parabola.
  (c) catenary.
  (d) hyperbola.

The answer is (b)

Problem–39

A cable carrying a constant uniform load per unit length along the length of the cable (e.g., a loose cable loaded by its own weight) will have the shape of a

  (a) straight line.
  (b) parabola.
  (c) catenary.
  (d) hyperbola.

The answer is (c)

Problem–40

Which of the following statements about two-dimensional mechanisms is not true?

(a) They are nonrigid structures.
(b) Relationships between forces can often be determined by statics.
(c) Several free body diagrams may be needed for complicated mechanisms.
(d) Generally, the resultant force on such a free body diagram will be in the direction of the member.

The answer is (d)

Problem–41

A statically indeterminant structure is one for which the equations of statics are not sufficient to determine all

(a) reactions.
(b) moments.
(c) internal stress.
(d) reactions, moments, and internal forces.

The answer is (d)

Problem–42

Which type of load is not resisted by a pinned joint?

(a) shear
(b) axial
(c) moment
(d) compression

The answer is (c)

Problem–43

The consistent deformation method of analyzing statically indeterminant beams and columns is based mainly on the fact that

(a) forces must be the same at a common point.
(b) deformations must be the same at a common point.
(c) moments must be the same at a common point.
(d) members cannot elongate or compress.

The answer is (b)

Problem–44

All of the following are methods applicable to the solution of simple statically indeterminant beams and trusses except the

(a) consistent deformation method.
(b) cut-and-sum method.
(c) superposition method.
(d) dummy unit load.

The answer is (b)

Problem–45

For a truss to be stable, all structural cells must be

(a) parallelograms.
(b) rectangles.
(c) squares.
(d) triangles.

The answer is (d)

Refer to the following illustration for Probs. 46 through 50.

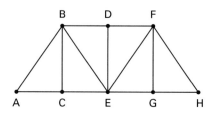

Problem–46

The end post of the bridge truss shown is part

(a) AC.
(b) AB.
(c) EF.
(d) G.

The answer is (b)

Problem–47

The joint of the bridge truss shown is part

  (a) AC.

  (b) BF.

  (c) FD.

  (d) B.

The answer is (d)

Problem–48

The web member of the bridge truss shown is part

  (a) ED.

  (b) D.

  (c) BF.

  (d) AB.

The answer is (a)

Problem–49

The upper chord of the bridge truss shown is part

  (a) BC.

  (b) FH.

  (c) BD.

  (d) BE.

The answer is (c)

Problem–50

The panel length of the bridge truss shown is part

  (a) AB.

  (b) ED.

  (c) BF.

  (d) CE.

The answer is (d)

Refer to the following illustration for Probs. 51 through 53.

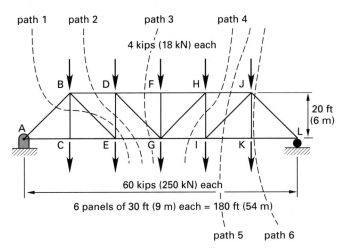

Problem–51

The fastest way to find the force through member DE is to use the

  (a) method of joints starting at the left end.

  (b) method of joints starting at the right end.

  (c) cut-and-sum method cutting through members along path 1.

  (d) cut-and-sum method cutting through members along path 2.

The answer is (c)

Problem–52

The fastest way to find the force through member HJ is to use the

  (a) method of joints starting at the left end.

  (b) method of joints starting at the right end.

  (c) method of sections cutting through members along path 4.

  (d) method of sections cutting through members along path 5.

The answer is (d)

Problem–53

The easiest way to calculate the load in member DE using the method of joints is starting at

    (a) A.
    (b) C.
    (c) D.
    (d) E.

The answer is (c)

Problem–54

The primary attribute of the cut-and-sum method for determining loads in statically determinant trusses is

    (a) it requires knowledge of reactions at all joints to the left of the member of interest.
    (b) it requires knowledge of reactions at all joints to the right of the member of interest.
    (c) it is strictly an application of the vertical equilibrium condition.
    (d) it is convenient when only a few truss members are known.

The answer is (c)

Problem–55

The primary attribute of the method of sections for determining loads in statically determinant trusses is

    (a) it requires knowledge of reactions at all joints to the left of the member of interest.
    (b) it requires knowledge of reactions at all joints to the right of the member of interest.
    (c) it is strictly an application of the vertical equilibrium condition.
    (d) it is convenient when only a few truss members are known.

The answer is (d)

Problem–56

The first step in finding the loads in any particular member of a statically determinant truss is to

    (a) decide whether or not to use the method of joints.
    (b) decide whether or not to use the cut-and-sum method.
    (c) decide whether or not to use the method of sections.
    (d) find the reactions.

The answer is (d)

Problem–57

For a statically determinant truss of two dimensions, the method of sections is limited to making a cut through the truss, which involves a maximum of how many unknown force members?

    (a) 1
    (b) 2
    (c) 3
    (d) 4

The answer is (c)

Problem–58

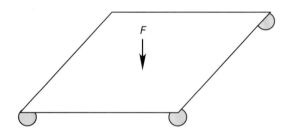

For the three-dimensional structure shown supported by ball bearings, how many reaction forces act at each corner?

    (a) 1
    (b) 2
    (c) 3
    (d) 4

The answer is (a)

Problem–59

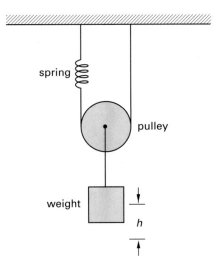

For the pulley system shown, if the weight drops a distance h, how far will the spring stretch?

(a) $\frac{1}{2}h$
(b) h
(c) $\frac{3}{2}h$
(d) $2h$

The answer is (d)

Refer to the following illustration for Probs. 60 and 61.

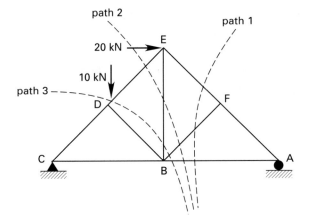

Problem–60

After finding the reaction at C, the quickest way to find the load in member DE is which of the following?

(a) Cut the truss along path 3 and apply the method of sections.
(b) Cut the truss along path 2 and apply the method of sections.
(c) Cut the truss along path 2 and apply the cut-and-sum method.
(d) Apply the method of joints starting at joint C.

The answer is (b)

Problem–61

The quickest way to find the force in member FB is to do which of the following?

(a) Cut the truss along path 1 and set the sum of moments about A to zero.
(b) Cut the truss along path 2 and set the sum of moments about A to zero.
(c) Cut the truss along path 3 and set the sum of moments about A to zero.
(d) Use the method of joints starting at point A.

The answer is (a)

Refer to the following illustration for Probs. 62 and 63.

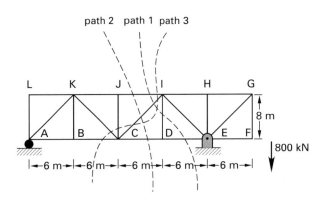

Problem—62

The quickest way to find the force through member IC is to do which of the following?

    (a) Find the reaction at E, cut the truss along path 1, and apply the cut-and-sum method.

    (b) Find the reaction at E, cut the truss along path 2, and apply the cut-and-sum method.

    (c) Find the reaction at E, cut the truss along path 3, and apply the cut-and-sum method.

    (d) Find the reaction at E, cut the truss along path 1, and apply the method of joints starting at A.

The answer is (b)

Problem—63

The quickest way to find the force through member BC is to do which of the following?

    (a) Cut through path 1, find F_{ka} from trigonometry, and then sum the forces at joint B.

    (b) Cut through path 2, find F_{ka} from trigonometry, and then sum the forces at joint B.

    (c) Cut through path 3, find F_{ka} from trigonometry, and then sum the forces at joint B.

    (d) Apply the method of joints starting at joint B.

The answer is (c)

Problem—64

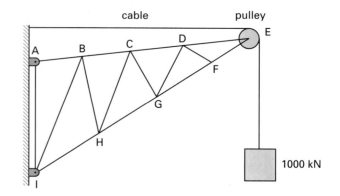

The force in the truss member AI closest to the wall is which of the following?

    (a) −250 kN

    (b) −100 kN

    (c) 0

    (d) 1000 kN

The answer is (c)

Problem—65

For statically indeterminant structures, the method of consistent deformations is based on the premise that

    (a) the energy (work done) by the deformations are equal.

    (b) the forces causing the deflections are equal.

    (c) the deflections for different members of the structure are equal.

    (d) the thermal deflections equal the mechanical deflections.

The answer is (c)

Chapter 3
Dynamics

Problem–1

The study of moving objects is called

(a) dynamics.
(b) statics.
(c) kinematics.
(d) kinetics.

The answer is (a)

Problem–2

The study of rigid bodies that are stationary is called

(a) dynamics.
(b) statics.
(c) kinematics.
(d) kinetics.

The answer is (b)

Problem–3

The field of dynamics that is the study of a body's motion independent of the forces on the body is called

(a) dynamics.
(b) kinematics.
(c) kinetics.
(d) indeterminant statics.

The answer is (b)

Problem–4

The study of the motion and the forces causing the motion of the body is called

(a) dynamics.
(b) kinematics.
(c) kinetics.
(d) indeterminant statics.

The answer is (c)

Problem–5

The number of coordinates required to completely specify the dynamic state of an object is the

(a) degree of static indeterminancy.
(b) degree of static determinancy.
(c) number of spatial coordinates.
(d) number of degrees of freedom.

The answer is (d)

Problem–6

Displacement in dynamics is

(a) the net change in a particle's position as determined from the position function.
(b) the accumulated length of the path traveled during all direction reversals.
(c) movement of particles in straight lines.
(d) motion in which acceleration increases or decreases linearly with time.

The answer is (a)

Problem–7

In dynamics, a linear system describes

(a) the accumulated length of the path traveled during all direction reversals.
(b) movement of particles in straight lines.
(c) uniform velocity.
(d) motion in which acceleration increases or decreases linearly with time.

The answer is (b)

Problem–8

Distance traveled in dynamics is

   (a) the net change in a particle's position as deter-
      mined from the position function.
   (b) the accumulated length of the path traveled dur-
      ing all direction reversals.
   (c) movement of particles in straight lines.
   (d) motion in which acceleration increases or de-
      creases linearly with time.

The answer is (b)

Problem–9

In dynamics, linear acceleration is

   (a) the net change in a particle's position as deter-
      mined from the position function.
   (b) movement of particles in straight lines.
   (c) uniform velocity.
   (d) motion in which acceleration increases or de-
      creases linearly with time.

The answer is (d)

Problem–10

In dynamics, uniform motion means

   (a) the accumulated length of the path traveled dur-
      ing all direction reversals.
   (b) movement of particles in straight lines.
   (c) uniform velocity.
   (d) motion in which acceleration increases or de-
      creases linearly with time.

The answer is (c)

Refer to the following illustrations for Probs. 11 and 12.

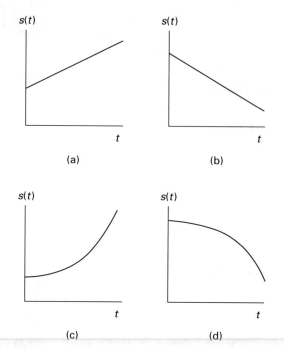

(a)   (b)

(c)   (d)

Problem–11

For a rigid body with a uniform positive acceleration, its
displacement is best described by which of the curves?

   (a) a
   (b) b
   (c) c
   (d) d

The answer is (c)

Problem–12

For a rigid body with uniform negative acceleration, its
displacement is described by which of the curves?

   (a) a
   (b) b
   (c) c
   (d) d

The answer is (d)

Problem–13

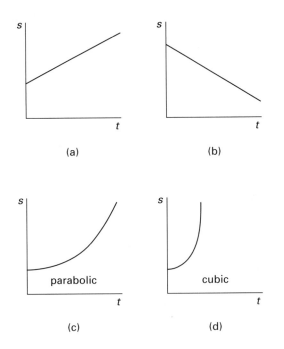

(a)

(b)

(c) parabolic

(d) cubic

For a body with linear increasing acceleration, its position is best given by which of the following?

(a) a
(b) b
(c) c
(d) d

The answer is (d)

Problem–14

For a projectile placed in motion by an initial impulse with both vertical and horizontal components, with no air drag, and experiencing only the downward body force of gravity, all of the following statements about its motion are true except

(a) The trajectory is parabolic.
(b) The impact velocity is equal to the initial velocity, v_0.
(c) The impact angle is equal to its launch angle ϕ.
(d) The range is maximum when the launch angle is 37°.

The answer is (d)

Problem–15

In dynamics, rotational motion is

(a) rotation of the body about its own axis.
(b) motion along any curved path.
(c) motion along any parabolic path.
(d) motion of a particle around a circular path.

The answer is (d)

Problem–16

Which of the following expressions does not correctly describe straight line motion with constant acceleration?

(a) $v = v_0 + a_0 t$
(b) $s = s_0 + v_0 t + \dfrac{a_0^2 t}{2}$
(c) $s = s_0 + \dfrac{v^2 - v_0^2}{2a_0}$
(d) $s = s_0 + a_0 t + \dfrac{v_0 t^2}{2}$

The answer is (d)

Problem–17

All of the following expressions correctly describe projectile motion in two dimensions except

(a) $v_x = v_{x0} = v \cos\theta$
(b) $v_y = v_{y0} - gt = v_0 \sin\theta - gt$
(c) $x = v_{x0}t = v_0 t \cos\theta$
(d) $x = x_0 + gt + v_{x0}\dfrac{t^2}{2}$

The answer is (d)

Problem–18

The acceleration produced when a particle moves away from its center of rotation is called the

(a) tangential acceleration.
(b) Coriolis acceleration.
(c) centripetal acceleration.
(d) centrifugal acceleration.

The answer is (b)

Problem—19

Acceleration toward the center of rotation is

(a) normal acceleration.
(b) Coriolis acceleration.
(c) centripetal acceleration.
(d) centrifugal acceleration.

The answer is (c)

Problem—20

The apparent force on a body directed away from the center of rotation is

(a) normal force.
(b) Coriolis force.
(c) centripetal force.
(d) centrifugal force.

The answer is (d)

Problem—21

The force of the plane reaction in friction calculations is

(a) tangential force.
(b) normal force.
(c) centripetal force.
(d) centrifugal force.

The answer is (b)

Problem—22

A body at rest on an inclined plane with coefficient of friction μ will not slip until the angle of the plane θ reaches

(a) $\tan\theta > \mu$.
(b) $\sin\theta > \mu$.
(c) $\cos\theta > \mu$.
(d) $(1 - \tan\theta) > \mu$.

The answer is (a)

Problem—23

The force with impending motion of a block on a plane in friction calculations is

(a) tangential force.
(b) normal force.
(c) centripetal force.
(d) centrifugal force.

The answer is (b)

Problem—24

The force that makes hurricanes flow counterclockwise in the northern hemisphere is

(a) gravity.
(b) tangential force.
(c) normal force.
(d) Coriolis force.

The answer is (d)

Problem—25

The description of motion of a particle with respect to the motion of another particle in motion is called

(a) linear motion.
(b) circular motion.
(c) relative motion.
(d) relativisitic motion.

The answer is (c)

Problem—26

The relative position of particle A with respect to particle B is best described by the equation

(a) $s_{B|A} = s_B + a_0 t$
(b) $s_{B|A} = s_B + s_A$
(c) $s_{B|A} = s_B - s_A$
(d) $s_{B|A} = s_B + v_0 t$

The answer is (c)

Problem–27

The instantaneous center of a rotating body can be found at the intersection of

(a) two known absolute velocity vectors.
(b) tangents to two known velocity vectors.
(c) perpendiculars to two known velocity vectors.
(d) a velocity vector and its normal.

The answer is (c)

Problem–28

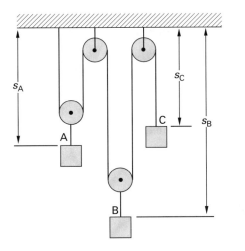

For the block and pulley system illustrated, which of the following is not true?

(a) $2s_A + 2s_B + s_C = $ constant
(b) Since the position of the nth block in an n-block system is determined when the remaining $n - 1$ positions are known, the number of degrees of freedom is equal to the number of blocks.
(c) $2v_A + 2v_B + v_C = 0$
(d) The movement, velocity, and acceleration of a block supported by two ropes are one-half the same quantities for a block supported by one rope.

The answer is (b)

Problem–29

All of the following are types of rigid body motion except

(a) pure translation. The orientation of the object is unchanged as its position changes.
(b) rotation about a fixed axis. All particles within the body move in ellipses, like the planets moving about the sun.
(c) general plane motion. The motion can be represented in two dimensions.
(d) motion about a fixed point. This describes any three-dimensional motion with one fixed point, such as a spinning top or truck-mounted crane.

The answer is (b)

Problem–30

The basis of the subject of kinetics is

(a) the second law of thermodynamics.
(b) Newton's laws.
(c) Fourier's laws.
(d) Le Châtelier's principle.

The answer is (b)

Problem–31

When a body remains in its displaced state after being moved from its original equilibrium position, the state which exists is called

(a) unstable equilibrium.
(b) stable equilibrium.
(c) neutral equilibrium.
(d) dynamic instability.

The answer is (c)

Problem-32

When a body returns to its original equilibrium position after being moved from it, the state which exists is called

  (a) unstable equilibrium.
  (b) stable equilibrium.
  (c) neutral equilibrium.
  (d) dynamic instability.

The answer is (b)

Problem-33

When a body moves away from its displaced state after being moved from its original equilibrium position, the state which exists is called

  (a) unstable equilibrium.
  (b) stable equilibrium.
  (c) neutral equilibrium.
  (d) dynamic instability.

The answer is (a)

Problem-34

All of the following are examples of external forces on a rigid body except

  (a) gravity.
  (b) electrostatics.
  (c) magnetic forces.
  (d) van der Waal's forces.

The answer is (d)

Problem-35

The law of conservation of momentum states that linear momentum is unchanged if the particle experiences

  (a) constant velocity.
  (b) constant mass.
  (c) no unbalanced forces.
  (d) constant acceleration.

The answer is (c)

Problem-36

When a ballistic pendulum is used to measure the velocity of a bullet on impact, what property is conserved?

  (a) kinetic energy
  (b) linear momentum
  (c) angular momentum
  (d) potential energy

The answer is (b)

Problem-37

To calculate the unknown velocity of a known mass with a ballistic pendulum, what two quantities are exchanged after impact?

  (a) linear momentum and angular momentum
  (b) kinetic energy and potential energy
  (c) linear momentum and potential energy
  (d) angular momentum and kinetic energy

The answer is (b)

Problem-38

The coefficient of restitution describes the behavior of what two quantities during the collision of billiard balls?

  (a) velocity ratios (before and after the collision)
  (b) momentum ratios (before and after the collision)
  (c) velocity differences (before and after the collision)
  (d) kinetic energy (before and after the collision)

The answer is (c)

Problem-39

In a perfectly elastic collision between two billiard balls, the coefficient of restitution has a value of

  (a) −1.
  (b) 0.
  (c) 1.
  (d) positive infinity.

The answer is (c)

Problem–40

In a perfectly inelastic collision between two billiard balls, the coefficient of restitution has a value of

    (a) negative infinity.
    (b) −1.
    (c) 0.
    (d) 1.

The answer is (c)

Problem–41

When a particle remains at a state of rest or continues to move with constant velocity, unless an unbalanced external force acts on it, the particle acts according to

    (a) Newton's first law of motion.
    (b) Newton's second law of motion.
    (c) Newton's third law of motion.
    (d) Einstein's theory of relativity.

The answer is (a)

Problem–42

If the resultant external force acting on a particle is zero, then the linear momentum p of the particle is

    (a) Newton's first law of motion.
    (b) Newton's second law of motion.
    (c) Newton's third law of motion.
    (d) Einstein's theory of relativity.

The answer is (a)

Problem–43

If the acceleration of a particle is directly proportional to the force acting on it and inversely proportional to the particle mass, it follows

    (a) Newton's first law of motion.
    (b) Newton's second law of motion.
    (c) Newton's third law of motion.
    (d) the first law of thermodynamics.

The answer is (b)

Problem–44

For every acting force between two bodies, there is an equal but opposite reacting force on the same line of action according to

    (a) Newton's first law of motion.
    (b) Newton's second law of motion.
    (c) Newton's third law of motion.
    (d) the first law of thermodynamics.

The answer is (c)

Problem–45

If the inertial force is included in the static equilibrium equation, then the body is in dynamic equilibrium. This is known as

    (a) Newton's second law of motion.
    (b) Newton's third law of motion.
    (c) Einstein's theory of relativity.
    (d) D'Alembert's principle.

The answer is (d)

Problem–46

The friction force that resists motion or impending motion is known most completely as

    (a) dynamic friction.
    (b) Coulomb friction.
    (c) fluid friction.
    (d) static, Coulomb, or fluid friction.

The answer is (d)

Problem–47

The friction force that resists motion when the body is already moving is known as

    (a) skin friction.
    (b) dynamic friction.
    (c) Coulomb friction.
    (d) fluid friction.

The answer is (b)

Problem-48

The relationship between dynamic and static friction is best described by the (in)equality

    (a) static friction $<<$ dynamic friction.
    (b) static friction $<$ dynamic friction.
    (c) static friction $=$ dynamic friction.
    (d) static friction $>$ dynamic friction.

The answer is (d)

Refer to the following illustration for Probs. 49 through 51.

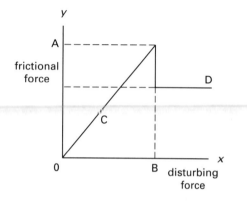

Problem-49

The frictional force during motion is best described by which parameter?

    (a) A
    (b) B
    (c) C
    (d) D

The answer is (d)

Problem-50

The frictional force prior to impending motion is best described by which parameter?

    (a) A
    (b) B
    (c) C
    (d) D

The answer is (c)

Problem-51

The disturbing force at impending motion is best described by which parameter?

    (a) A
    (b) B
    (c) C
    (d) D

The answer is (b)

Problem-52

The attractive gravitational force between two masses is directly proportional to

    (a) each of the masses.
    (b) Newton's gravitational force G.
    (c) the product of the two masses.
    (d) the square of the distance between the two masses.

The answer is (d)

Problem-53

Planets orbiting the sun sweep out equal areas in equal times according to

    (a) the first law of thermodynamics.
    (b) one of Kepler's laws of planetary motion.
    (c) Newton's second law of motion.
    (d) Charles' law.

The answer is (b)

Problem—54

If the oscillatory motion of a body about an equilibrium point is the result of a disturbing force being applied once, and then removed, then the motion is known as

(a) forced vibration.
(b) natural (free) vibration.
(c) damped vibration.
(d) undamped vibration.

The answer is (b)

Problem—55

If the oscillatory motion of a body about an equilibrium point is the result of a disturbing force applied repeatedly, the motion is known as

(a) forced vibration.
(b) natural (free) vibration.
(c) damped vibration.
(d) undamped vibration.

The answer is (a)

Problem—56

If the oscillatory motion of a body about an equilibrium point is reduced by friction, the motion is known as

(a) forced vibration.
(b) natural (free) vibration.
(c) damped vibration.
(d) undamped vibration.

The answer is (c)

Problem—57

The equation describing the motion of an undamped oscillation $\Sigma F = 0: -ma - kx = 0$ stems from

(a) Boyle's law.
(b) Newton's third law.
(c) Newton's second law.
(d) Newton's first law.

The answer is (c)

Problem—58

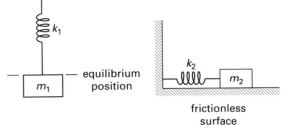

If $m_1 = m_2$ and $k_1 = k_2$, the relationship between the periods T_1 and T_2 of the two springs is which of the following?

(a) $T_1 << T_2$
(b) $T_1 < T_2$
(c) $T_1 = T_2$
(d) $T_1 > T_2$

The answer is (c)

Problem—59

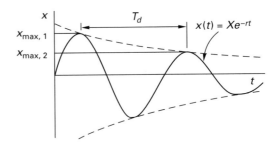

The oscillation shown is which of the following?

(a) underdamped
(b) undamped
(c) overdamped
(d) critically damped

The answer is (a)

Problem–60

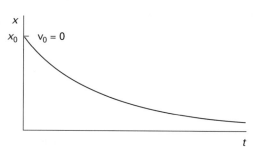

The oscillation shown is which of the following?

   (a) underdamped
   (b) undamped
   (c) overdamped
   (d) critically damped

The answer is (c)

Problem–61

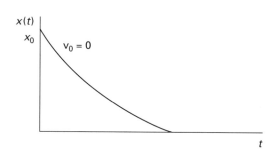

The oscillation shown is which of the following?

   (a) underdamped
   (b) overdamped
   (c) critically damped
   (d) negatively damped

The answer is (c)

Problem–62

In real oscillating systems, damping is

   (a) always present.
   (b) sometimes present.
   (c) never present.
   (d) always neglected.

The answer is (a)

Problem–63

For the free vibration of a simple spring and mass system initially displaced from equilibrium, the equation of motion is

$$ma = -kx$$

Its solution is

$$x(t) = x_0 \cos \omega t + \left(\frac{v_0}{\omega}\right) \sin \omega t$$

An alternative form of the solution is

   (a) $x(t) = A \sin^2 (\omega t + \phi)$.
   (b) $x(t) = A \cos (\omega t - \phi)$.
   (c) $x(t) = A \tan (\omega t - \phi)$.
   (d) $x(t) = A \cos^{-1} (\omega t + \phi)$.

The answer is (b)

Problem–64

What is the natural frequency ω of an oscillating system?

   (a) $\dfrac{k}{m}$

   (b) $\sqrt{\dfrac{k}{m}}$

   (c) $\dfrac{m}{k}$

   (d) $\sqrt{\dfrac{m}{k}}$

The answer is (b)

Problem–65

For the free vibration of a simple spring and mass system initially displaced from equilibrium, the equation of motion is
$$ma = -kx$$

Its solution is
$$x(t) = x_0 \cos \omega t + \left(\frac{v_0}{\omega}\right) \sin \omega t = A \cos (\omega t - \phi)$$

What does the coefficient A in the solution equal?

(a) $\sqrt{x_0^2 + \left(\frac{v_0}{\omega}\right)^2}$

(b) v_{max}

(c) A_{max}

(d) $\sqrt{\frac{k}{m}}$

The answer is (a)

Problem–66

For the free vibration of a simple spring and mass system initially displaced from equilibrium, the equation of motion is
$$ma = -kx$$

Its solution is
$$x(t) = x_0 \cos \omega t + \left(\frac{v_0}{\omega}\right) \sin \omega t = A \cos (\omega t - \phi)$$

The phase angle ϕ is given by

(a) $\arctan \left(\frac{v_0}{\omega x_0}\right)$.

(b) $\arctan \left(\frac{v_0 \omega}{x_0}\right)$.

(c) $\arctan \left(\frac{v_0^2}{x_0 \omega}\right)$.

(d) $\tan \left(\frac{v_0}{x_0 \omega}\right)$.

The answer is (a)

Problem–67

For the free vibration of a simple spring and mass system initially displaced from equilibrium, the equation of motion is
$$ma = -kx$$

Its solution is
$$x(t) = x_0 \cos \omega t + \left(\frac{v_0}{\omega}\right) \sin \omega t = A \cos (\omega t - \phi)$$

What does the coefficient A equal?

(a) x_{max}
(b) v_{max}
(c) A_{max}
(d) $\sqrt{\frac{k}{m}}$

The answer is (a)

Problem–68

For the free vibration of a simple spring and mass system initially displaced from equilibrium, the equation of motion is
$$ma = -kx$$

Its solution is
$$x(t) = x_0 \cos \omega t + \left(\frac{v_0}{\omega}\right) \sin \omega t = A \cos (\omega t - \phi)$$

What does the product $A\omega$ equal?

(a) x_{max}
(b) v_{max}
(c) A_{max}
(d) $\sqrt{\frac{k}{m}}$

The answer is (b)

Problem–69

For the free vibration of a simple spring and mass system initially displaced from equilibrium, the equation of motion is

$$ma = -kx$$

Its solution is

$$x(t) = x_0 \cos \omega t + \left(\frac{v_0}{\omega}\right) \sin \omega t = A \cos(\omega t - \phi)$$

What does the product $A\omega^2$ equal?

(a) x_{max}
(b) v_{max}
(c) A_{max}
(d) $\sqrt{\dfrac{k}{m}}$

The answer is (c)

Problem–70

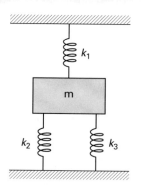

Three springs in parallel have an equivalent spring constant equal to which of the following?

(a) $k_{\text{eq}} = k_1 + k_2 + k_3$
(b) $k_{\text{eq}} = k_1 - k_2 - k_3$
(c) $k_{\text{eq}} = -k_1 + k_2 + k_3$
(d) $k_{\text{eq}} = \dfrac{1}{k_1} + \dfrac{1}{k_2} + \dfrac{1}{k_3}$

The answer is (a)

Problem–71

For natural, undamped vibrations, the natural period of oscillation is affected by

(a) initial position.
(b) initial velocity.
(c) both initial position and initial velocity.
(d) neither initial position nor initial velocity.

The answer is (d)

Problem–72

For natural, undamped vibrations, the amplitude of the oscillation is affected by

(a) initial position.
(b) initial velocity.
(c) both initial position and initial velocity.
(d) neither initial position nor initial velocity.

The answer is (c)

Problem–73

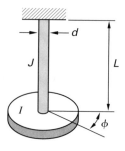

The device illustrated is which of the following?

(a) free pendulum
(b) torsional pendulum
(c) forced pendulum
(d) spring mass system

The answer is (b)

Problem-74

The natural frequency of a torsional pendulum is given by which of the following?

(a) $\dfrac{k_r}{I}$

(b) $\dfrac{I}{k_r}$

(c) $\sqrt{\dfrac{k_r}{I}}$

(d) $\sqrt{\dfrac{I}{k_r}}$

The answer is (c)

Problem-75

In the solution of the equation of motion for a torsional pendulum,

$$\phi(t) = \phi_0 \cos \omega t + \left(\frac{\omega_0}{\omega}\right) \sin \omega t = A \cos(\omega t - \theta)$$

The coefficient A is given by which of the following?

(a) $\sqrt{\phi_0^2 - \left(\dfrac{\omega_0}{\omega}\right)^2}$

(b) $A \cos(\omega t - \theta)$

(c) $\sqrt{\phi_0^2 + \left(\dfrac{\omega_0}{\omega}\right)^2}$

(d) $\sqrt{\phi_0^2 - \left[\tan^{-1}\left(\dfrac{\omega_0}{\omega}\right)^2\right]}$

The answer is (c)

Problem-76

In the solution of the equation of motion for a torsional pendulum, the phase angle θ is given by

(a) $\arcsin\left(\dfrac{\omega_0}{\omega \phi_0}\right)$.

(b) $\phi_{\max}$.

(c) $\arctan\left(\dfrac{\omega_0}{\omega \phi_0}\right)$.

(d) $\omega_{\max}$.

The answer is (c)

Problem-77

In the solution of the equation of motion for a torsional pendulum, the product $A\omega$ is given by

(a) $\phi_{\max}$.

(b) $\arctan\left(\dfrac{\omega_0}{\omega \phi_0}\right)$.

(c) $\omega_{\max}$.

(d) $\alpha_{\max}$.

The answer is (c)

Problem-78

In the solution of the equation of motion for a torsional pendulum, the product $A\omega^2$ is given by

(a) $\phi_{\max}$.

(b) $\arctan\left(\dfrac{\omega_0}{\omega \phi_0}\right)$.

(c) $\omega_{\max}$.

(d) $\alpha_{\max}$.

The answer is (d)

Problem-79

The equation $m\ddot{x} = -kx$ describes the motion for

(a) damped, forced vibration.
(b) free vibration.
(c) damped, free vibration.
(d) undamped, forced vibration.

The answer is (b)

Problem-80

The equation $m\ddot{x} = -kx + F_0 \cos \omega_f t$ describes the motion for

(a) damped, forced vibration.
(b) free vibration.
(c) damped, free vibration.
(d) undamped, forced vibration.

The answer is (d)

Problem–81

The equation $m\ddot{x} = -kx - C\dot{x}$ describes the motion for

   (a) damped, forced vibration.
   (b) free vibration.
   (c) damped, free vibration.
   (d) undamped, forced vibration.

The answer is (c)

Problem–82

The equation $m\ddot{x} = -kx - C\dot{x} + F_0 \cos \omega_f t$ describes the motion for

   (a) damped, forced vibration.
   (b) free rotation.
   (c) free vibration.
   (d) damped, free vibration.

The answer is (a)

Problem–83

The equation $-k_r \phi = I\ddot{\phi}$ describes the motion for

   (a) damped, forced vibration.
   (b) free rotation.
   (c) free vibration.
   (d) undamped, forced vibration.

The answer is (b)

Problem–84

In vibrating systems, the ratio of the steady state vibration magnitude D to the pseudostatic deflection, F_0/k is known as the

   (a) damping factor or damping ratio.
   (b) reactive factor.
   (c) magnification factor.
   (d) critical coefficient of viscous damping.

The answer is (c)

Problem–85

In damped free vibrations, the ratio $C/2\sqrt{mk}$ where C is the coefficient of viscous damping, and m and k are the mass and spring constants respectively, is called

   (a) damping factor or damping ratio.
   (b) reactive factor.
   (c) magnification factor.
   (d) critical coefficient of viscous damping.

The answer is (a)

Problem–86

The denominator of the expression for the damping factor $\zeta = C/2\sqrt{mk}$ is called the

   (a) damping factor or damping ratio.
   (b) reactive factor.
   (c) magnification factor.
   (d) critical damping coefficient.

The answer is (d)

Problem–87

In damped free vibrations, for a damping factor $\zeta < 1$, that is, $C < C_{\text{critical}}$, the motion is

   (a) overdamped.
   (b) the same as natural vibration.
   (c) oscillatory with diminishing magnitude.
   (d) critically damped.

The answer is (c)

Problem–88

In damped free vibrations, the boundary between underdamped and damped vibrations (e.g., between the motions shown in Probs. 59 and 60) is marked by

   (a) $\zeta < 1$.
   (b) $\zeta = 1$.
   (c) $\zeta > 1$.
   (d) $\zeta \gg 1$.

The answer is (b)

Problem—89

For a vibrating system, resonance occurs when the natural frequency ω of the system and the forcing frequency ω_f have which of the following relationships?

  (a) $\omega << \omega_f$
  (b) $\omega < \omega_f$
  (c) $\omega = \omega_f$
  (d) $\omega > \omega_f$

The answer is (c)

Problem—90

For a vibrating system with no damping at resonance, the magnification factor becomes

  (a) very small.
  (b) zero.
  (c) very large.
  (d) infinite.

The answer is (d)

Problem—91

The physical significance of an infinite magnification factor (resonance) is that amplitude of vibration becomes

  (a) zero.
  (b) small.
  (c) very large.
  (d) infinite.

The answer is (d)

Problem—92

The natural logarithm of the ratio of the amplitude of two successive underdamped free oscillations is known as the

  (a) free logarithm.
  (b) natural logarithm.
  (c) logarithmic mean difference.
  (d) logarithmic decrement.

The answer is (d)

Problem—93

In damped free vibration, when the system returns to its static position with no overshoot past equilibrium, the system is said to be

  (a) underdamped.
  (b) overdamped.
  (c) resonant.
  (d) critically damped.

The answer is (d)

Problem—94

In a critically damped, free vibrational system, the roots of the characteristic equation to the differential equation are

  (a) unequal and imaginary.
  (b) equal and real.
  (c) unequal and real.
  (d) zero.

The answer is (b)

Problem—95

Impulse is

  (a) force times distance.
  (b) a vector quantity equal to the change in velocity.
  (c) a scalar.
  (d) a vector quantity equal to the change in momentum.

The answer is (d)

Problem—96

The impulse-momentum principle may be applied in each of the following systems except

  (a) determining the thrust in a jet engine.
  (b) determining the recoil force of a cannon.
  (c) determining the work done against friction.
  (d) finding the force of an impact.

The answer is (c)

Problem—97

Rolling resistance is due to

   (a) friction between body and surface.
   (b) change in linear momentum.
   (c) change in angular momentum.
   (d) deformation of the rolling body and the surface.

The answer is (d)

Problem—98

The capacity of a mass to do work is represented by its

   (a) temperature.
   (b) volume.
   (c) momentum.
   (d) energy.

The answer is (d)

Problem—99

Total energy of a body can be calculated from its mass and

   (a) temperature.
   (b) specific volume.
   (c) momentum.
   (d) specific energy.

The answer is (d)

Problem—100

Energy cannot be created or destroyed according to the

   (a) law of conservation of momentum.
   (b) first law of thermodynamics.
   (c) law of conservation of energy.
   (d) Newton's second law.

The answer is (c)

Problem—101

The act of changing the energy of a particle, body, or system is

   (a) thermodynamics.
   (b) fluid mechanics.
   (c) work.
   (d) kinetic energy.

The answer is (c)

Problem—102

Work is a(n)

   (a) unsigned scalar quantity.
   (b) signed scalar quantity.
   (c) unsigned vector quantity.
   (d) signed vector quantity.

The answer is (b)

Problem—103

All of the following units are acceptable for quantifying work except

   (a) foot-pounds.
   (b) watt-seconds.
   (c) joules.
   (d) dynes.

The answer is (d)

Problem—104

How is work in a linear system calculated?

   (a) $\int \mathbf{F} \cdot \mathbf{ds}$

   (b) $\dfrac{kx^2}{2}$

   (c) $\dfrac{\rho g h}{g_c}$

   (d) $\frac{1}{2}m\mathbf{v}^2$

The answer is (a)

Problem–105

How is work in a rotational system calculated?

(a) $\dfrac{kx^2}{2}$

(b) $\displaystyle\int T \cdot d\theta$

(c) ρgh

(d) $\frac{1}{2}Iw^2$

The answer is (b)

Problem–106

Work can be converted to all of the following forms except

(a) kinetic energy.
(b) potential energy.
(c) thermal energy.
(d) horsepower.

The answer is (d)

Problem–107

Power is

(a) force integrated over distance.
(b) force integrated over time.
(c) force per unit time.
(d) work per unit time.

The answer is (d)

Problem–108

All of the following are acceptable units for expressing power except

(a) ft-lb.
(b) hp.
(c) W.
(d) J/s.

The answer is (a)

Refer to the following illustration for Probs. 109 through 113.

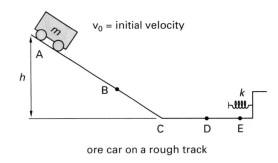

ore car on a rough track

Problem–109

At point A, what is the total energy of the cart?

(a) $E_A = \frac{1}{2}mv_0^2$

(b) $E_A = mgh$

(c) $E_A = mgh + \frac{1}{2}mv_0^2$

(d) $E_A = mgh - \frac{1}{2}mv_0^2$

The answer is (c)

Problem–110

At point B, the energy of the cart is

(a) $E_B = E_A - F_f s_{AB}$

(b) $E_B = mgh - F_f s_{AB}$

(c) $E_B = \frac{1}{2}mv_0^2 - F_f s_{AB}$

(d) $E_B = \frac{1}{2}mv_0^2 + F_f s_{AB}$

The answer is (a)

Problem–111

At point C in the movement of the cart down the slope toward the spring, how is the velocity of the cart found?

(a) $\frac{1}{2}mv^2 = mgh - \frac{1}{2}mv_0^2 - F_f s_{AC}$

(b) $\frac{1}{2}mv^2 = mgh + \frac{1}{2}mv_0^2 + F_f s_{AC}$

(c) $\frac{1}{2}mv^2 = mgh + \frac{1}{2}mv_0^2 - F_f s_{AC}$

(d) $mv = mg + mv_0 - F_f$

The answer is (c)

Problem–112

At point D, what is the kinetic energy of the cart?

(a) $\frac{1}{2}mv^2 = mgh - \frac{1}{2}mv_0^2 - F_f s_{AD}$

(b) $\frac{1}{2}mv^2 = mgh + \frac{1}{2}mv_0^2 - F_f s_{AD} - \frac{1}{2}kx^2$

(c) $\frac{1}{2}mv^2 = \frac{1}{2}mv_0^2 - F_f s_{AD}$

(d) $\frac{1}{2}mv^2 = mgh + \frac{1}{2}mv_0^2 - F_f s_{AD}$

The answer is (d)

Problem–113

At point E, the compression of the spring, x, may be found from

(a) $kx = mg - F_f$

(b) $kx^2 = mgh - F_f s_{AE}$

(c) $\frac{1}{2}kx^2 = \frac{1}{2}mv_0^2 - F_f s_{AE}$

(d) $\frac{1}{2}kx^2 = mgh + \frac{1}{2}mv_0^2 - F_f(s_{AE} + x)$

The answer is (d)

Problem–114

Work done against friction can be expressed as

(a) $F_f s$.

(b) $mg(h_2 - h_1)$.

(c) $\frac{1}{2}mv^2$.

(d) $\frac{1}{2}k(x_2^2 - x_1^2)$.

The answer is (a)

Problem–115

Potential energy can be expressed as

(a) $F_f s$.

(b) $mg(h_2 - h_1)$.

(c) $\frac{1}{2}mv^2$.

(d) $\frac{1}{2}k(x_2^2 - x_1^2)$.

The answer is (b)

Problem–116

Kinetic energy in translation can be expressed as

(a) $F_f s$.

(b) $mg(h_2 - h_1)$.

(c) $\frac{1}{2}mv^2$.

(d) $\frac{1}{2}k(x_2^2 - x_1^2)$.

The answer is (c)

Problem–117

Kinetic energy in rotation can be expressed as

(a) $\frac{1}{2}I\omega^2$.

(b) $mg(h_2 - h_1)$.

(c) $\frac{1}{2}mv^2$.

(d) $\frac{1}{2}k(x_2^2 - x_1^2)$.

The answer is (a)

Problem–118

Work done against a spring can be expressed as

(a) $F_f s$.

(b) $mg(h_2 - h_1)$.

(c) $\frac{1}{2}mv^2$.

(d) $\frac{1}{2}k(x_2^2 - x_1^2)$.

The answer is (d)

Chapter 4
Mechanics of Materials

Problem–1

The amount of force required to cause a unit deformation (displacement)—often called the spring constant—is the

    (a) modulus of elasticity.
    (b) normal stress.
    (c) shear stress.
    (d) stiffness.

The answer is (d)

Problem–2

When more than one spring or resisting member shares the load, the relative stiffness is the

    (a) modulus of elasticity.
    (b) normal stress.
    (c) stiffness.
    (d) rigidity.

The answer is (d)

Problem–3

The units of relative stiffness or rigidity are

    (a) newtons.
    (b) newtons/meter.
    (c) newton-meters.
    (d) dimensionless.

The answer is (d)

Problem–4

Individual rigidity values

    (a) have no significance.
    (b) indicate how much stronger one member is than another.
    (c) are significant only for ductile materials.
    (d) are significant only for brittle materials.

The answer is (a)

Problem–5

The ratio of two rigidities

    (a) have no significance.
    (b) indicate how much stronger one member is than another.
    (c) are significant only for ductile materials.
    (d) are significant only for brittle materials.

The answer is (b)

Problem–6

The sum of the strains in the three orthogonal directions $(\epsilon_x, \epsilon_y, \epsilon_z)$ in accordance with Poisson's ratio is

    (a) Hooke's law.
    (b) the modulus of elasticity.
    (c) dilation.
    (d) the shear modulus.

The answer is (c)

Problem–7

The constant of proportionality relating the linear expansion of an object to its change in temperature is called its

    (a) modulus of elasticity.
    (b) shear modulus.
    (c) bulk modulus.
    (d) coefficient of linear expansion.

The answer is (d)

Problem–8

The constant of proportionality relating the volumetric expansion of an object to its change in temperature is called its

    (a) modulus of elasticity.
    (b) bulk modulus.
    (c) coefficient of linear expansion.
    (d) coefficient of volumetric expansion.

The answer is (d)

Problem–9

The thermal stress in a bar that is allowed to expand freely when heated to a higher temperature is

    (a) large and positive.
    (b) small and positive.
    (c) zero.
    (d) small and negative.

The answer is (c)

Problem–10

Very low values of the coefficient of thermal expansion lead to

    (a) very low values of thermal expansion.
    (b) no thermal expansion.
    (c) high values of thermal expansion.
    (d) very high values of thermal expansion.

The answer is (a)

Problem–11

Materials with low values of thermal expansion are desirable for

    (a) Pyrex glassware.
    (b) barometers.
    (c) thermometers.
    (d) railroad rails.

The answer is (a)

Problem–12

All of the following shapes can lead to stress concentrations except

    (a) stepped shafts.
    (b) plates with holes.
    (c) gently tapered parts with large fillets and radii.
    (d) shafts and keyways.

The answer is (c)

Problem–13

Stress concentration factors are not normally applied to any of the following configurations except

    (a) static loading of ductile materials.
    (b) applications with local yielding around discontinuities.
    (c) applications where stresses are kept low by design.
    (d) long grooves where the ratio of groove radius to shaft diameter is small.

The answer is (d)

Problem–14

When applied normal and shear stresses are resolved in such a manner that the shear stresses vanish (go to zero), the resulting normal stresses are called the

    (a) main stresses.
    (b) normal stresses.
    (c) extreme shear stresses.
    (d) principal stresses.

The answer is (d)

Problem–15

When applied normal and shear stresses are resolved in such a manner that the shear stresses vanish (go to zero), the resulting normal stresses at that point are

(a) resolvable to shear stresses.
(b) negligible.
(c) the maximum and minimum stresses acting in any direction.
(d) important only in compression.

The answer is (c)

Problem–16

If the applied stress is known as a function of θ, then the principal stresses may be found by the mathematical process of

(a) reduction by partial fractions.
(b) addition.
(c) differentiation.
(d) integration.

The answer is (c)

Refer to the following illustration for Probs. 17 through 21.

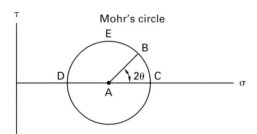

Problem–17

On Mohr's circle for stress (which relates the applied normal stresses σ_x and σ_y and the shear stress τ), the average normal stress, $(1/2)(\sigma_x + \sigma_y)$, is shown by point

(a) A.
(b) B.
(c) C.
(d) D.

The answer is (a)

Problem–18

On Mohr's circle for stress the maximum principal stress is shown by point

(a) A.
(b) B.
(c) C.
(d) D.

The answer is (c)

Problem–19

On Mohr's circle for stress the minimum principal stress is shown by point

(a) A.
(b) B.
(c) C.
(d) D.

The answer is (d)

Problem–20

On Mohr's circle for stress the plane of the principal stresses is indicated by the angle

(a) $-\theta$.
(b) -2θ.
(c) 2θ.
(d) θ.

The answer is (d)

Problem–21

On Mohr's circle for stress the extreme shear stresses are indicated by the ordinate at point

(a) E.
(b) B.
(c) C.
(d) D.

The answer is (a)

Problem–22

Young's modulus is

(a) the ratio of lateral strain to axial strain.
(b) the percentage deformation under tension.
(c) a measure of force across a cross section.
(d) the ratio of stress to strain.

The answer is (d)

Problem–23

If Young's modulus for a steel cable is 10×10^{10} N/m^2, the actual change in length of a 100 m cable under a stress of 10^7 Pa is

(a) 1 mm.
(b) 1 cm.
(c) 10 cm.
(d) 1 m.

The answer is (c)

Refer to the following illustration for Probs. 24 through 26.

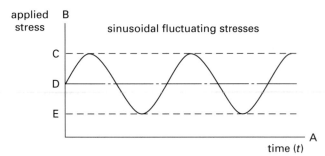

Problem–24

In the illustration of alternating stresses shown, point C indicates the

(a) mean stress.
(b) maximum stress.
(c) applied stress.
(d) minimum stress.

The answer is (b)

Problem–25

In the illustration of alternating stresses shown, point E indicates the

(a) mean stress.
(b) maximum stress.
(c) applied stress.
(d) minimum stress.

The answer is (d)

Problem–26

In the illustration of alternating stresses shown, point D indicates the

(a) mean stress.
(b) maximum stress.
(c) applied stress.
(d) minimum stress.

The answer is (a)

Refer to the following illustration for Probs. 27 through 33.

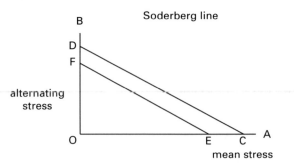

Problem–27

The graphical criterion for acceptable design (or unacceptable failure) due to alternating stresses is called the

  (a) shear diagram.
  (b) Soderberg line.
  (c) Goodman line.
  (d) Mohr's circle.

The answer is (b)

Problem–28

The failure line is given by

  (a) DF.
  (b) EC.
  (c) DC.
  (d) FE.

The answer is (c)

Problem–29

The allowable stress line is given by

  (a) DF.
  (b) EC.
  (c) DC.
  (d) FE.

The answer is (d)

Problem–30

The maximum value of alternating stress on the failure line is given by

  (a) the endurance limit, S_e, at point D.
  (b) the yield strength, S_{yt}, at point C.
  (c) the endurance limit, S_e, divided by a safety factor, FS, at point F.
  (d) the yield strength, S_{yt}, divided by a safety factor, FS, at point E.

The answer is (a)

Problem–31

The maximum value of alternating stress on the allowable stress line is given by

  (a) the endurance limit, S_e, at point D.
  (b) the yield strength, S_{yt}, at point C.
  (c) the endurance limit, S_e, divided by a safety factor, FS, at point F.
  (d) the yield strength, S_{yt}, divided by a safety factor, FS, at point E.

The answer is (c)

Problem–32

The minimum value of alternating stress on the allowable stress line is given by the mean stress value of

  (a) the endurance limit, S_e, at point D.
  (b) the yield strength, S_{yt}, at point C.
  (c) the endurance limit, S_e, divided by a safety factor, FS, at point F.
  (d) the yield strength, S_{yt}, divided by a safety factor, FS, at point E.

The answer is (d)

Problem–33

In the illustration shown, the acceptable design region is the area

  (a) above DC.
  (b) below DC.
  (c) below FE.
  (d) between DC and FE.

The answer is (c)

Refer to the following illustration for Probs. 34 through 39.

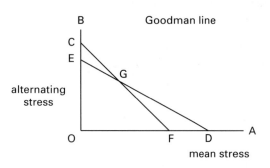

Problem–34

Another graphical criterion for acceptable design (or unacceptable failure) of alternating stresses is called the

(a) shear diagram.
(b) Goodman line.
(c) Mohr's circle.
(d) moment diagram.

The answer is (b)

Problem–35

The failure line is given by

(a) CF.
(b) ED.
(c) EGF.
(d) CGD.

The answer is (c)

Problem–36

The maximum value of alternating stress on the failure line is given by

(a) the endurance limit, S_e, at point E.
(b) the yield strength, S_{yt}, at point C.
(c) the yield strength, S_{yt}, divided by a safety factor, FS, at point F.
(d) the ultimate strength, S_{ut}, at point D.

The answer is (a)

Problem–37

Point C on the figure corresponds to

(a) the endurance limit, S_e.
(b) the yield strength, S_{yt}.
(c) the endurance limit, S_e, divided by a safety factor, FS.
(d) the ultimate strength, S_{ut}.

The answer is (b)

Problem–38

Point D on the figure corresponds to

(a) the endurance limit, S_e.
(b) the yield strength, S_{yt}.
(c) the yield strength, S_{yt}, divided by a safety factor, FS.
(d) the ultimate strength, S_{ut}.

The answer is (d)

Problem–39

The acceptable design region of the figure is indicated by the area bounded by

(a) OCF.
(b) OED.
(c) OCD.
(d) OEGF.

The answer is (d)

Problem–40

The disadvantage of using the Goodman design criterion instead of the Soderberg design criterion is that it

(a) applies only for brittle materials.
(b) is the more conservative of fluctuating stress theories.
(c) is acceptable for metals only below their critical temperature.
(d) requires the material's ultimate strength to be known.

The answer is (d)

Problem-41

Like the Soderberg criterion, the Goodman criterion should be used with all of the following materials except

(a) steel.
(b) aluminum.
(c) titanium.
(d) cast iron.

The answer is (d)

Problem-42

The shear at a point on a loaded beam is the

(a) algebraic sum of all the beam's reaction loads acting between the investigation point and one of the ends.
(b) sum of all vertical forces acting between the investigation point and one of the ends.
(c) sum of all vertical stresses acting between the investigation point and one of the ends.
(d) sum of the beam's reaction moments.

The answer is (b)

Problem-43

The moment at a point on a loaded beam is

(a) the algebraic sum of all moments and couples acting between the investigation point and one of the ends.
(b) the algebraic sum of all the beam's reaction loads acting between the investigation point and one of the ends.
(c) the sum of all vertical stresses acting between the investigation point and one of the ends.
(d) the sum of the beam's reaction moments.

The answer is (a)

Problem-44

Which of the following statements about shear diagrams is not true?

(a) The shear at any point is equal to the sum of the loads and reactions from the point to the left end.
(b) The magnitude of the shear at any point is equal to the slope of the moment line at that point.
(c) Loads and reactions upward are positive.
(d) The shear diagram is curved and concave downward over uniformly distributed loads.

The answer is (d)

Problem-45

Which of the following statements about shear diagrams is not true?

(a) The shear at any point is equal to the sum of the loads and reactions from the point to the left end.
(b) The magnitude of the shear at any point is equal to the integral of the moment line at that point.
(c) Loads and reactions upward are positive.
(d) The shear diagram is straight and sloping over uniformly distributed loads.

The answer is (b)

Problem-46

Which of the following statements about moment diagrams is not true?

(a) Clockwise moments about the point are negative. The right-hand rule should be used to determine positive moments.
(b) The magnitude of the moment at any point is equal to the area under the shear line up to that point. This is equivalent to the integral of the shear function: $M = \int V\,dx$.
(c) The maximum moment occurs where the shear is zero.
(d) The moment diagram is straight and sloping between concentrated loads.

The answer is (a)

Problem–47

Which of the following statements about moment diagrams is not true?

(a) Clockwise moments about the point are positive. The left-hand rule should be used to determine positive moments.

(b) The magnitude of the moment at any point is equal to the area under the shear line up to that point. This is equivalent to the integral of the shear function: $M = \int V\,dx$.

(c) The moment diagram is curved (parabolic downward) over uniformly distributed loads.

(d) The moment diagram is straight and sloping between concentrated loads.

The answer is (c)

Problem–48

In calculations using standard structural shapes of I-beams, it is assumed that the shear stress is carried by the

(a) entire cross-sectional area.

(b) web only.

(c) top flange only when in compression.

(d) flanges only.

The answer is (b)

Problem–49

In a beam of rectangular cross section, the maximum shear stress when related to the average shear stress is

(a) 150% higher.

(b) 50% higher.

(c) 33% higher.

(d) 50% lower.

The answer is (b)

Problem–50

In a beam of circular cross section, the maximum shear stress when related to the average shear stress is

(a) 150% higher.

(b) 50% higher.

(c) 33% higher.

(d) 50% lower.

The answer is (c)

Problem–51

The bending stress (flexural stress) of a beam is really

(a) a normal stress whose source is bending.

(b) a normal stress whose source is shear.

(c) due to internal pressure.

(d) a newly recognized stress from internal bending of the cross section.

The answer is (a)

Problem–52

A positive moment on a beam causes what kind of stress in the upper fibers of the beam?

(a) compression

(b) tension

(c) shear

(d) fatigue

The answer is (a)

Problem–53

The maximum bending stress a beam experiences occurs at the

(a) neutral axis.

(b) surface furthest removed from the neutral axis.

(c) end of the beam.

(d) middle of the beam.

The answer is (b)

Problem–54

When a load is applied to a beam axially at the centroid of the cross section, the beam is said to be loaded

   (a) eccentrically.
   (b) concentrically.
   (c) dynamically.
   (d) pseudostatically.

The answer is (b)

Problem–55

When a load is applied to a beam axially but not at the centroid of the cross section, the beam is said to be loaded

   (a) eccentrically.
   (b) concentrically.
   (c) dynamically.
   (d) pseudostatically.

The answer is (a)

Problem–56

When an axial stress is applied eccentrically to a beam, the combined stress due to compression (tension) and bending may be found by

   (a) simple addition.
   (b) combined stress theory.
   (c) Mohr's circle.
   (d) finite element analysis.

The answer is (a)

Problem–57

Which of the following expressions relating beam deflections, slopes, moments, and the like, is not correct?

   (a) $y = $ deflection

   (b) $\dfrac{d^3y}{dx^3} = \dfrac{V(x)}{EI}$

   (c) $V(x) = \displaystyle\int M(x)dx$

   (d) $\dfrac{d^2y}{dx^2} = \dfrac{M(x)}{EI}$

The answer is (c)

Problem–58

The strain energy method of calculating beam deflection at a point of load application equates the total internal strain energy to the external

   (a) deflection.
   (b) force.
   (c) moment.
   (d) work.

The answer is (d)

Problem–59

In the moment area method for computing beam deflection, the area of the moment diagram (between two points on the beam) divided by EI gives the

   (a) deflection between tangents at the two points.
   (b) energy absorbed between points.
   (c) angle between tangents at the two points.
   (d) shear between the two points.

The answer is (c)

Problem–60

In the moment area method for computing beam deflection, the statical moment of the bending between two points divided by EI

$$\int \frac{xM(x)}{EI}\,dx$$

gives the

   (a) deflection of the beam.
   (b) energy absorbed between points.
   (c) angle between tangents at the two points.
   (d) shear between the two points.

The answer is (a)

Refer to the following illustration for Probs. 61 and 62.

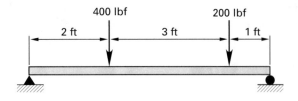

Problem–61

What is the maximum shear?

(a) 100 lbf
(b) 200 lbf
(c) 300 lbf
(d) 400 lbf

The answer is (c)

Problem–62

What is the maximum bending moment?

(a) 300 ft-lbf
(b) 600 ft-lbf
(c) 900 ft-lbf
(d) 1800 ft-lbf

The answer is (b)

Problem–63

The conjugate beam method of calculating beam deflection allows the deflection to be found by drawing

(a) shear diagrams.
(b) shear stress diagrams.
(c) moment diagrams.
(d) shear and moment diagrams.

The answer is (c)

Problem–64

One of the great advantages to using the conjugate beam method to compute beam deflections is that the method can handle beams

(a) of varying cross section.
(b) with different materials.
(c) with varying cross section and different materials.
(d) with two built-in ends.

The answer is (c)

Problem–65

One of the great disadvantages to using the conjugate beam method to compute beam deflections is that the method cannot handle beams

(a) of varying cross section.
(b) with different materials.
(c) with varying cross section and different materials.
(d) with two built-in ends.

The answer is (d)

Problem–66

The easiest, most straight-forward method for determining beam deflection is the

(a) double integration method.
(b) moment area method.
(c) strain energy method.
(d) table look-up method.

The answer is (d)

Problem–67

The easiest, most straight-forward method for determining beam deflections when multiple loads act simultaneously is superposition and the

(a) double integration method.
(b) moment area method.
(c) strain energy method.
(d) table look-up method.

The answer is (d)

Problem–68

The method of superposition for calculating beam deflections due to multiple loads is valid as long as

(a) none of the deflections are excessive.
(b) stresses are less than the yield strength.
(c) none of the deflections are excessive and stresses are less than the yield strength.
(d) deflections are negligibly small.

The answer is (c)

Problem–69

All of the following are feasible methods for calculating truss deflections of statically indeterminant trusses except the

(a) strain energy method.
(b) virtual work method.
(c) unit load method.
(d) table look-up method.

The answer is (d)

Problem–70

All of the following are recognized types of beam failure except

(a) excessive deflection.
(b) local buckling.
(c) rotational buckling.
(d) longitudinal buckling.

The answer is (d)

Refer to the following illustrations for Probs. 71 through 75.

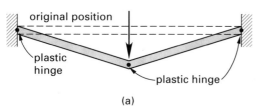

(a)

(b)

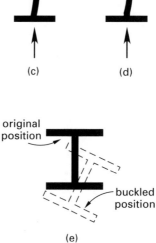

(c) (d)

(e)

Problem–71

An example of vertical buckling is shown by illustration

(a) a.
(b) b.
(c) c.
(d) d.

The answer is (c)

Problem–72

An example of web crippling is shown by illustration

- (a) a.
- (b) b.
- (c) c.
- (d) d.

The answer is (d)

Problem–73

An example of lateral buckling is shown by illustration

- (a) a.
- (b) b.
- (c) d.
- (d) e.

The answer is (d)

Problem–74

An example of rotation is shown by illustration

- (a) a.
- (b) b.
- (c) c.
- (d) e.

The answer is (a)

Problem–75

An example of column buckling is shown by illustration

- (a) a.
- (b) b.
- (c) c.
- (d) d.

The answer is (b)

Problem–76

What is the material stress for ductile materials, in the calculation of allowable stress from the formula?

$$\text{allowable stress} = \frac{\text{material stress}}{\text{factor of safety}}$$

- (a) yield strength
- (b) ultimate strength
- (c) endurance strength
- (d) mean stress

The answer is (a)

Problem–77

What is the material stress for brittle materials, in the calculation of allowable stress from the formula?

$$\text{allowable stress} = \frac{\text{material stress}}{\text{factor of safety}}$$

- (a) yield strength
- (b) ultimate strength
- (c) endurance strength
- (d) mean stress

The answer is (b)

Problem–78

Failure of slender columns occurs when

- (a) applied stress exceeds yield strength.
- (b) applied stress exceeds ultimate strength.
- (c) allowable stress exceeds the factor of safety.
- (d) the load exceeds the critical or Euler load (and the column buckles sideways).

The answer is (d)

Problem–79

The length, l, of a column divided by r is one of the terms in the equation for the buckling of a column under compression loads. What does r stand for in the l/r ratio?

  (a) radius of the column
  (b) radius of gyration
  (c) moment of inertia
  (d) slenderness

The answer is (b)

Problem–80

The buckling load of intermediate columns (too tall to be piers, but too short to be slender columns) can be determined by

  (a) the Euler load with fixed-end conditions.
  (b) computing allowable stress from the yield strength.
  (c) computing allowable stress from the ultimate strength.
  (d) the secant formula.

The answer is (d)

Problem–81

Which of the following gives the equivalent (composite) spring constant for a number of springs in series?

  (a) $\dfrac{1}{k_{\text{eq}}} = \dfrac{1}{k_1} + \dfrac{1}{k_2} + \ldots + \dfrac{1}{k_n}$

  (b) $k_{\text{eq}} = k_1 + k_2 + \ldots + k_n$

  (c) $k_{\text{eq}} = \dfrac{1}{n} \displaystyle\sum_{i=1}^{n} k_i$

  (d) $k_{\text{eq}} = \displaystyle\prod_{i=1}^{n} k_i$

The answer is (a)

Problem–82

Which of the following gives the equivalent (composite) spring constant for a number of springs in parallel?

  (a) $\dfrac{1}{k_{\text{eq}}} = \dfrac{1}{k_1} + \dfrac{1}{k_2} + \ldots + \dfrac{1}{k_n}$

  (b) $k_{\text{eq}} = k_1 + k_2 + \ldots + k_n$

  (c) $k_{\text{eq}} = \dfrac{1}{n} \displaystyle\sum_{i=1}^{n} k_i$

  (d) $k_{\text{eq}} = \displaystyle\prod_{i=1}^{n} k_i$

The answer is (b)

Problem–83

Which of the following gives the hoop stress (σ_h) of a thin-walled cylindrical tank?

  (a) $\sigma_h = \dfrac{pr}{t}$

  (b) $\sigma_h = \dfrac{pr}{2t}$

  (c) $\dfrac{t}{2r} < 0.1$

  (d) $\sigma_r = \dfrac{p_i r_i^2}{(r_o + r_i)t}$

The answer is (a)

Problem–84

Which of the following gives the hoop stress (σ_h) of a thin-walled spherical tank?

  (a) $\sigma_h = \dfrac{pr}{t}$

  (b) $\sigma_h = \dfrac{pr}{2t}$

  (c) $\dfrac{t}{2r} < 0.1$

  (d) $\sigma_r = \dfrac{p_i r_i^2}{(r_o + r_i)t}$

The answer is (b)

Problem–85

Which of the following gives the hoop stress (σ_h) of a thick-walled cylindrical tank?

(a) $\sigma_h = \dfrac{pr}{t}$

(b) $\sigma_h = \dfrac{pr}{2t}$

(c) $\dfrac{t}{2r} < 0.1$

(d) $\sigma_r = \dfrac{p_i r_i^2}{(r_o + r_i)t}$

The answer is (d)

Problem–86

The maximum shear stress is experienced by which of the following bolt(s)?

(a) A
(b) B
(c) B and C
(d) C and A

The answer is (c)

Refer to the following illustrations for Probs. 87 through 90.

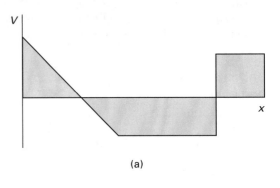

(a)

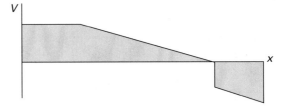

(b)

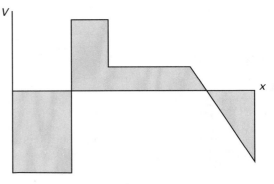

(c)

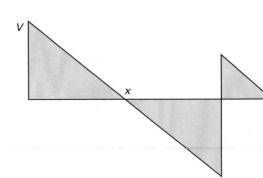

(d)

Problem–87

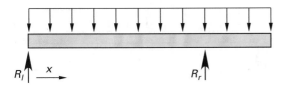

The shear diagram for the beam loading shown is most nearly

- (a) a.
- (b) b.
- (c) c.
- (d) d.

The answer is (d)

Problem–88

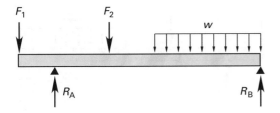

The shear diagram for the beam loading shown is most nearly

- (a) a.
- (b) b.
- (c) c.
- (d) d.

The answer is (c)

Problem–89

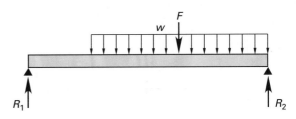

The shear diagram for the beam loading shown is most nearly

- (a) a.
- (b) b.
- (c) c.
- (d) d.

The answer is (b)

Problem–90

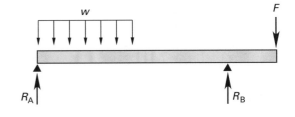

The shear diagram for the beam loading shown is most nearly

- (a) a.
- (b) b.
- (c) c.
- (d) d.

The answer is (a)

Refer to the following illustrations for Probs. 91 through 94.

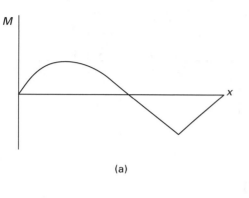

(a)

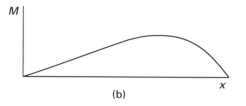

(b)

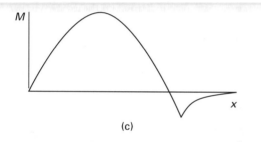

(c)

(d)

Problem–91

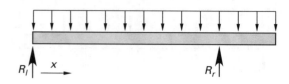

The moment diagram for the beam loading shown is most nearly

    (a) a.
    (b) b.
    (c) c.
    (d) d.

The answer is (c)

Problem–92

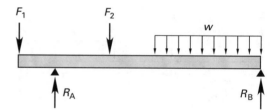

The moment diagram for the beam loading shown is most nearly

    (a) a.
    (b) b.
    (c) c.
    (d) d.

The answer is (d)

Problem—93

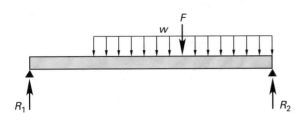

The moment diagram for the beam loading shown is most nearly

(a) a.
(b) b.
(c) c.
(d) d.

The answer is (b)

Problem—94

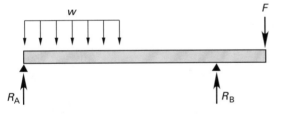

The moment diagram for the beam loading shown is most nearly

(a) a.
(b) b.
(c) c.
(d) d.

The answer is (a)

Problem—95

A good mnemonic device for remembering whether fibers in the upper surface of a beam experiencing a positive moment are in compression or tension is, "A positive moment makes a beam

(a) cry."
(b) smile."
(c) frown."
(d) neutral."

The answer is (b)

Chapter 5

Fluid Mechanics

Problem–1

Liquids and gases take the following characteristic(s) of their containers.

    (a) volume
    (b) shape
    (c) shape and volume
    (d) neither shape nor volume

The answer is (b)

Problem–2

For computational convenience, fluids are usually classed as

    (a) rotational or irrotational.
    (b) real or ideal.
    (c) laminar or turbulent.
    (d) Newtonian or dilatent.

The answer is (b)

Problem–3

Which of the following statements about a Newtonian fluid is most accurate?

    (a) Shear stress is proportional to strain.
    (b) Viscosity is zero.
    (c) Shear stress is multivalued.
    (d) Shear stress is proportional to rate of strain.

The answer is (d)

Problem–4

Which of the following is not a characteristic of real fluids?

    (a) finite viscosity
    (b) non-uniform velocity distributions
    (c) compressibility
    (d) experience of eddy currents and turbulence

The answer is (d)

Problem–5

Which of the following is not the mass density of water?

    (a) 62.4 lbm/ft^3
    (b) 100 kg/m^3
    (c) 1 g/cm^3
    (d) 1 kg/L

The answer is (b)

Problem–6

Which of the following is not a characteristic of fluid pressure?

    (a) It is the same in all directions at a point in the fluid.
    (b) It acts normal to a surface.
    (c) It is a shear stress.
    (d) It is linear with depth.

The answer is (c)

Problem–7

Which of the following statements about gauge pressure is most correct? Gauge pressures are measured relative to

- (a) atmospheric pressure.
- (b) a vacuum.
- (c) each other.
- (d) the surface.

The answer is (a)

Problem–8

Gauge pressure and absolute pressure differ from each other by

- (a) the system of units.
- (b) atmospheric pressure.
- (c) the size of the gauge.
- (d) nothing—they mean the same thing.

The answer is (b)

Problem–9

Each of the following are correct values of standard atmospheric pressure except

- (a) 1.000 atm.
- (b) 14.692 psia.
- (c) 760 torr.
- (d) 1013 mm Hg.

The answer is (d)

Problem–10

When a pump is operating at a vacuum of 4 in Hg, which of the following is not correct?

- (a) The pressure is 25.92 in Hg.
- (b) The pressure is 10.721 psia.
- (c) The pressure is 158.4 torr.
- (d) The pressure is 0.8663 atm.

The answer is (c)

Problem–11

All of the following are properties of an ideal gas except

- (a) density.
- (b) pressure.
- (c) viscosity.
- (d) temperature.

The answer is (c)

Problem–12

Which of the following is not the universal gas constant?

- (a) 1545 ft-lb/lbmol-°R
- (b) 8.314 H/mol·K
- (c) 8314 kJ/mol·K
- (d) 8.314 kJ/kmol·K

The answer is (c)

Problem–13

The following are all commonly quoted values of standard temperature and pressure except

- (a) 32°F and 14.696 psia.
- (b) 273.15K and 101.325 kPa.
- (c) 0°C and 760 mm Hg.
- (d) 0°F and 29.92 in Hg.

The answer is (d)

Problem–14

Kinematic and dynamic viscosity vary from each other only by a factor equal to the

- (a) fluid density.
- (b) temperature.
- (c) pressure.
- (d) specific gas constant.

The answer is (a)

Problem–15

All of the following dimensionless parameters are applicable to fluid flow problems except the

(a) Reynolds number.
(b) Froude number.
(c) Mach number.
(d) Biot number.

The answer is (d)

Problem–16

Compressibility of a fluid relates the fractional change in fluid volume per unit change in fluid

(a) temperature.
(b) density.
(c) pressure.
(d) viscosity.

The answer is (c)

Problem–17

The speed of sound in a fluid is most closely related to all of the following properties except

(a) compressibility.
(b) density.
(c) bulk modulus.
(d) thermal conductivity.

The answer is (d)

Problem–18

The term *subsonic flow* refers to a flowing gas with a speed

(a) less than the local speed of sound.
(b) equal to the speed of sound.
(c) greater than the speed of sound.
(d) much greater than the speed of sound.

The answer is (a)

Problem–19

The variation of pressure in an isobaric process is

(a) linear with temperature.
(b) described by the perfect gas law.
(c) inversely proportional to temperature.
(d) zero.

The answer is (d)

Problem–20

All of the following can be characteristics of fluids except

(a) kinematic viscosity.
(b) surface tension.
(c) bulk modulus.
(d) hysteresis.

The answer is (d)

Problem–21

The statement that "the hydrostatic pressure a fluid exerts on an immersed object or on container walls is a function only of fluid depth" is

(a) the perfect gas law.
(b) D'Alembert's paradox.
(c) the hydrostatic paradox.
(d) Boyle's law.

The answer is (c)

Problem–22

Bernoulli's equation is a(n)

(a) momentum equation.
(b) conservation of energy equation.
(c) conservation of mass equation.
(d) equation of state.

The answer is (b)

Problem–23

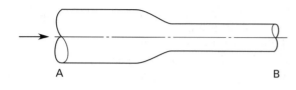

A　　　　　　　　　　　B

What is the ratio of the velocity at B to that at A in the round pipe shown?

(a) $v_B > v_A$
(b) $v_B \gg v_A$
(c) $v_B = v_A$
(d) $v_B < v_A$

The answer is (a)

Problem–24

The pressure at a given depth due to several immiscible liquids is

(a) the average of the individual pressures.
(b) the sum of the individual pressures.
(c) independent of the individual pressures.
(d) unknown.

The answer is (b)

Problem–25

The relationship between pressure and altitude in the atmosphere is given by the

(a) perfect gas law.
(b) conservation of mass.
(c) barometric height relationship.
(d) first law of thermodynamics.

The answer is (c)

Problem–26

The fact that the buoyant force on a floating object is equal to the weight of the water displaced is

(a) Bernoulli's law.
(b) Archimedes' principle.
(c) the law of diminishing returns.
(d) the conservation of mass.

The answer is (b)

Problem–27

Which of the following terms does not appear in the steady flow energy equation (the extended Bernoulli equation)?

(a) kinetic energy
(b) potential energy
(c) friction losses
(d) hysteresis losses

The answer is (d)

Problem–28

A pitot tube can be used to measure fluid velocity as described by the Bernoulli equation and the relationship between

(a) kinetic energy and static pressure.
(b) fluid pressure and height of the fluid.
(c) fluid pressure and impact energy.
(d) pressure and momentum.

The answer is (a)

Problem–29

The difference between stagnation pressure and total pressure is

(a) due to height difference.
(b) due to fluid kinetic energy.
(c) none—the terms are interchangeable.
(d) important only in supersonic flow.

The answer is (c)

Problem–30

Fully turbulent flow in a pipe is characterized by all of the following except

  (a) a parabolic velocity profile.
  (b) a momentum exchange due to fluid masses rather than molecules.
  (c) a maximum velocity at the fluid center line.
  (d) a $\frac{1}{7}$ velocity profile.

The answer is (a)

Problem–31

The laminar friction factor of fluid flowing through a pipe is a function of all the following except

  (a) fluid velocity.
  (b) pipe diameter.
  (c) pipe roughness.
  (d) Reynolds number.

The answer is (c)

Problem–32

The stream function is a useful parameter in describing

  (a) the conservation of mass.
  (b) the conservation of momentum.
  (c) the conservation of energy.
  (d) the equation of state.

The answer is (a)

Problem–33

Which of the following is the correct statement for the conservation of mass for incompressible flow?

  (a) $\frac{p}{\rho} + \frac{v^2}{2} = \text{constant}$
  (b) $vA = \text{constant}$
  (c) $p = \rho v^2 = \text{constant}$
  (d) $u = \frac{\partial \psi}{\partial y}$

The answer is (b)

Problem–34

The study of the practical laws of fluid flow and the resistance of open pipes and channels is the purview of

  (a) fluid mechanics.
  (b) hydraulics.
  (c) aerodynamics.
  (d) thermodynamics.

The answer is (b)

Problem–35

The most common method for calculating frictional energy loss for laminar flowing fluids in noncircular pipes is

  (a) the Darcy equation.
  (b) the Hagan-Poiseville equation.
  (c) the Hazen-Williams equation.
  (d) the Swamee-Jain equation.

The answer is (a)

Problem–36

What is the most common method for calculating frictional energy loss for laminar flowing fluids in noncircular pipes?

  (a) $h_f = \dfrac{fLv^2}{2Dg}$
  (b) $E_f = \dfrac{32\mu vL}{D^2\rho}$
  (c) $h_f = \dfrac{3.022v^{1.85}L}{C^{1.85}D^{1.165}}$
  (d) $\dfrac{1}{\sqrt{f}} = 2.0\log_{10}\left(\text{Re}\sqrt{f} - 0.80\right)$

The answer is (a)

Problem–37

A circular pipe of radius 4 ft is flowing half-full. What is its hydraulic radius?

$A = \frac{\pi c^2}{2} = 8\pi$

$C = 2\pi r = 8\pi$

(a) 1 ft

(b) $\frac{\pi}{2}$ ft

(c) 2 ft

(d) 4 ft

The answer is (a)

Problem–38

A canal of rectangular cross section is 8 ft wide and water is flowing 4 ft deep. What is its hydraulic radius?

(a) 2 ft

(b) 4 ft

(c) 8 ft

(d) 16 ft

The answer is (a)

Problem–39

The parameter f in the expression for head loss is

(a) the fraction of flow that is totally turbulent.

(b) the Darcy friction factor.

(c) the height of the roughness scale in turbulent flow.

(d) the static coefficient of friction.

The answer is (b)

Problem–40

What is the friction factor for laminar flow?

(a) $f = \dfrac{0.316}{\mathrm{Re}^{0.25}}$

(b) $\dfrac{1}{\sqrt{f}} = 2.0 \log_{10}\left(\mathrm{Re}\sqrt{f} - 0.80\right)$

(c) $f - \dfrac{64}{\mathrm{Re}}$

(d) $\dfrac{1}{\sqrt{f}} = 1.74 - 2.0 \log_{10}\left(\dfrac{2\epsilon}{D}\right)$

The answer is (c)

Problem–41

Friction factors for both laminar and turbulent flows can be found plotted in a

(a) steam table.

(b) psychrometric chart.

(c) Moody diagram.

(d) Mollier diagram.

The answer is (c)

Problem–42

The characteristic length of the Reynolds number used to calculate the friction factor in noncircular full running pipes is based on the

(a) run length.

(b) pipe diameter.

(c) hydraulic diameter (the equivalent diameter).

(d) wetted circumference.

The answer is (c)

Problem–43

The hydraulic radius of a noncircular pipe is

(a) the square root of the flow area.

(b) the ratio of the flow area to the wetted perimeter.

(c) the radius of a pipe of equivalent area.

(d) none of the above.

The answer is (b)

Problem–44

The Darcy equation can be used for all liquids and flows except

(a) water.

(b) alcohol.

(c) gasoline.

(d) air flowing supersonically.

The answer is (d)

Problem–45

The Hazen-Williams formula for head loss due to friction is based upon

(a) rigorous mathematical derivation.
(b) empirical data.
(c) semi-empirical analysis.
(d) serendipity.

The answer is (b)

Problem–46

The extended Bernoulli equation includes all of the following terms except

(a) potential energy.
(b) kinetic energy.
(c) nuclear energy.
(d) friction losses.

The answer is (c)

Problem–47

The hydraulic grade line of a pipeline denotes which of the following?

(a) total energy
(b) pressure energy
(c) potential energy
(d) the sum of pressure energy and potential energy

The answer is (d)

Problem–48

The energy grade line of a pipeline denotes which of the following?

(a) total energy
(b) pressure energy
(c) potential energy
(d) the sum of pressure energy and potential energy

The answer is (a)

Problem–49

The presence of friction in the energy grade line will always cause the line to slope

(a) down in the direction of the flow.
(b) up in the direction of the flow.
(c) level (no slope).
(d) There is no effect of friction on the energy grade line.

The answer is (a)

Problem–50

The presence of friction in the hydraulic grade line will always cause the line to slope

(a) down in the direction of the flow.
(b) up in the direction of the flow.
(c) level (no slope).
(d) There is no effect of friction on the hydraulic grade line.

The answer is (a)

Problem–51

The presence of a minor loss in the energy grade line will cause the line to slope

(a) down in the direction of the flow.
(b) up in the direction of the flow.
(c) vertically down.
(d) There is no effect of friction on the energy grade line.

The answer is (a)

Problem–52

The presence of a sudden expansion in the hydraulic grade line will cause the line to slope

(a) down in the direction of the flow.
(b) up in the direction of the flow.
(c) vertically up.
(d) level (no slope change).

The answer is (c)

Problem–53

The difference between the energy grade line and the hydraulic grade line is

(a) potential energy.
(b) pressure energy.
(c) kinetic energy.
(d) friction losses.

The answer is (c)

Problem–54

The difference between the $EGL_{friction}$ and the $EGL_{frictionless}$ is

(a) pressure energy.
(b) kinetic energy.
(c) friction losses.
(d) friction and minor losses.

The answer is (d)

Problem–55

The amount of energy actually entering the fluid from a pump is the

(a) brake horsepower.
(b) hydraulic horsepower.
(c) theoretical horsepower.
(d) hydraulic, theoretical, or water horsepower.

The answer is (d)

Problem–56

The primary purpose of a pump in a fluid loop is to

(a) add energy to the flow.
(b) add mass to the flow.
(c) extract energy from the flow.
(d) none of the above

The answer is (a)

Problem–57

The primary purpose of a turbine in a fluid loop is to

(a) add energy to the flow.
(b) add mass to the flow.
(c) extract energy from the flow.
(d) none of the above

The answer is (c)

Problem–58

A vena contracta in a fluid jet issuing through a hole in a plate is located approximately

(a) 10 diameters downstream of the hole.
(b) at the jet's minimum diameter.
(c) at the orifice minimum diameter.
(d) at the orifice maximum diameter.

The answer is (b)

Problem–59

Orifice coefficients are used to determine

(a) energy losses.
(b) energy gains.
(c) mass losses.
(d) energy losses and mass gains.

The answer is (a)

Problem–60

The coefficient of velocity is the ratio of the

(a) area of a vena contracta to the orifice area.
(b) actual discharge to the theoretical discharge.
(c) actual discharge velocity to the theoretical discharge velocity.
(d) effective head to the actual head.

The answer is (c)

Problem–61

The coefficient of contraction is the ratio of the

(a) area of a vena contracta to the orifice area.
(b) actual discharge to the theoretical discharge.
(c) actual velocity to the theoretical velocity.
(d) effective head to the actual head.

The answer is (a)

Problem–62

The coefficient of discharge is the ratio of the

(a) area of a vena contracta to the orifice area.
(b) actual discharge to the theoretical discharge.
(c) actual velocity to the theoretical velocity.
(d) effective head to the actual head.

The answer is (b)

Problem–63

The coefficient of velocity is equal to the

(a) product of the coefficient of discharge and the coefficient of contraction.
(b) actual velocity divided by the theoretical velocity.
(c) sum of the coefficient of discharge and the coefficient of contraction.
(d) difference of the coefficient of discharge and the coefficient of contraction.

The answer is (b)

Problem–64

Discharge losses through orifices are due to

(a) friction losses.
(b) minor losses.
(c) both friction and minor losses.
(d) pressure losses.

The answer is (c)

Problem–65

Which of the following is not a similarity between a submerged culvert and a siphon?

(a) They both operate full.
(b) Torricelli's equation holds.
(c) Both can experience entrance and exit losses.
(d) In both, the water flows downhill.

The answer is (b)

Problem–66

In parallel pipe systems originating and terminating in common junctions,

(a) mass flows through each branch are equal.
(b) pressure drops through each branch are equal.
(c) lengths of each branch are equal.
(d) flow areas of each branch are equal.

The answer is (b)

Problem–67

Flows through multiloop systems may be computed by

(a) any closed-form solution of simultaneous equations.
(b) the Hardy-Cross method.
(c) trial and error.
(d) all of the above

The answer is (d)

Problem–68

Flow measuring devices include all of the following except

(a) venturi meters.
(b) static pressure probes.
(c) turbine and propeller meters.
(d) magnetic dynamometers.

The answer is (d)

Problem–69

Flow measuring devices include all of the following except

   (a) orifice plate meters.
   (b) hot-wire anemometers.
   (c) magnetic flow meters.
   (d) mercury barometers.

The answer is (d)

Problem–70

Flow measuring devices include all of the following except

   (a) flow nozzles.
   (b) venturi area meters.
   (c) pitot tubes.
   (d) precision tachometers.

The answer is (d)

Problem–71

The following are all examples of indirect (secondary) measurements to measure flow rates using obstruction meters except

   (a) variable area meters.
   (b) venturi meters.
   (c) volume tanks.
   (d) flow nozzles.

The answer is (c)

Problem–72

The following are all examples of indirect (secondary) measurements to measure flow rates using velocity meters except

   (a) pitot static meters.
   (b) static pressure probes.
   (c) weight and mass scales.
   (d) direction-sensing probes.

The answer is (c)

Problem–73

The following are all examples of indirect (secondary) miscellaneous methods to measure flow except

   (a) turbine and propeller meters.
   (b) magnetic flow meters.
   (c) positive displacement meters.
   (d) hot-wire anemometers.

The answer is (c)

Problem–74

In series pipe systems, all of the following parameters vary from section to section except

   (a) pressure drop.
   (b) friction loss.
   (c) head loss.
   (d) mass flow.

The answer is (d)

Problem–75

Venturi meters, pitot static gauges, orifice meters, flow nozzles, and differential manometers all depend upon a relationship between

   (a) flow velocity and friction.
   (b) flow velocity and pressure.
   (c) friction and pressure.
   (d) pressure and mass flow.

The answer is (b)

Problem–76

Given the relationships (1), (2), (3), and (4), which of the following combinations would best describe the relevant equations for compressible flow of an ideal gas?

(1) $p = \rho RT$

(2) $p = ZRT$

(3) $\rho vA = \text{constant}$

(4) Mollier Chart

  (a) (1) and (3)
  (b) (2) and (3)
  (c) (1), (2), and (3)
  (d) (2) and (4)

The answer is (a)

Problem–77

The coefficient of velocity, C_v, accounts for the

  (a) effects on the flow area of contraction, friction, and turbulence.
  (b) small effect of friction and turbulence of the orifice.
  (c) changes in diameters of a converging pipe.
  (d) effects of compressibility.

The answer is (b)

Problem–78

Expansion factors take into account the

  (a) area of the vena contracta.
  (b) small effect of friction and turbulence of the orifice.
  (c) changes in diameters of a converging pipe.
  (d) effects of compressibility.

The answer is (d)

Problem–79

In fluid flow, linear momentum is

  (a) a vector quantity equal to the product of mass and velocity.
  (b) a scalar quantity equal to the product of mass and velocity.
  (c) a scalar quantity equal to the product of force and length of time it is applied.
  (d) the change in impulse.

The answer is (a)

Problem–80

All of the following fluid phenomena are based on the force momentum principle of a flowing fluid except

  (a) turbines.
  (b) Pelton wheels.
  (c) diesel automobile engines.
  (d) jet engines.

The answer is (c)

Problem–81

The matching of scale model and full-scale results for a fluid dynamic phenomena with a free surface requires equality of

  (a) Reynolds number.
  (b) Weber number.
  (c) Froude number.
  (d) Cauchy number.

The answer is (c)

Problem–82

The matching of scale model and full-scale results for a fluid dynamic phenomena involving compressible fluids requires equality of

  (a) Reynolds number.
  (b) Froude number.
  (c) Cauchy number.
  (d) Mach number.

The answer is (d)

Problem–83

The matching of scale model and full-scale prototype results for a fluid dynamic phenomena involving surface tension requires equality of

(a) Reynolds number.
(b) Weber number.
(c) Froude number.
(d) Cauchy number.

The answer is (b)

Problem–84

The matching of scale model and full-scale results for a fluid dynamic phenomena involving a fully submerged body requires equality of

(a) Reynolds number.
(b) Weber number.
(c) Froude number.
(d) Mach number.

The answer is (a)

Problem–85

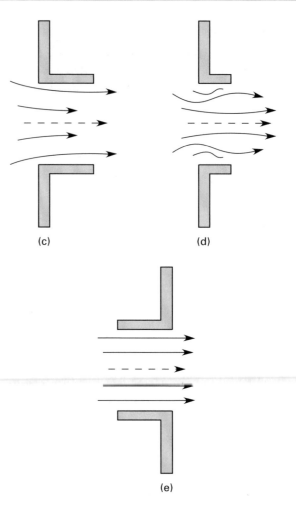

(c)   (d)

(e)

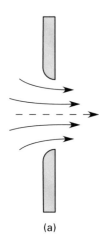

(a)

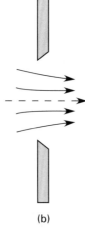

(b)

Consider the orifices illustrated. Rank them in order of decreasing loss coefficient—that is, in order from largest to smallest loss.

(a) e, d, c, a, b
(b) e, b, d, c, a
(c) c, e, b, d, a
(d) d, c, a, b, e

The answer is (b)

Problem–86

The water hammer phenomenon is primarily what kind of fluid mechanics?

  (a) static (a phenomenon independent of time)
  (b) dynamic (a time-dependent phenomenon)
  (c) compressible
  (d) incompressible

The answer is (b)

Problem–87

All of the following are forms of drag on a body moving through a fluid except

  (a) skin friction.
  (b) wake drag.
  (c) profile drag.
  (d) d'Alembert's paradox drag.

The answer is (d)

Problem–88

The magnitude of the drag coefficient of a sphere in water is dependent upon all of the following except

  (a) fluid density.
  (b) fluid velocity.
  (c) units of measure (SI or English Engineering System).
  (d) drag force.

The answer is (c)

Problem–89

The fact that there is no resistance to bodies moving through an ideal (nonviscous) fluid is known as

  (a) Reynold's analogy.
  (b) d'Alembert's paradox.
  (c) Newton's second law.
  (d) the second law of thermodynamics.

The answer is (b)

Problem–90

The fact that a fluid's velocity increases as the cross-sectional area of the pipe through which it flows decreases is due to

  (a) Bernoulli's equation.
  (b) the continuity equation.
  (c) the momentum equation.
  (d) the perfect gas law.

The answer is (b)

Problem–91

Which of the following is a dimensionless grouping for the energy extracted from a flow by a turbine (where P = power, Q = volumetric flow rate, γ = specific weight, H = head)?

  (a) $P\gamma Q H$

  (b) $\dfrac{\gamma Q}{PH}$

  (c) $\dfrac{P}{Q\gamma H}$

  (d) $\dfrac{Q\gamma}{PH}$

The answer is (c)

Problem–92

When a falling object reaches a speed at which the drag force equals its weight, it has achieved

  (a) Mach one.
  (b) a laminar boundary layer.
  (c) a turbulent boundary layer.
  (d) terminal velocity.

The answer is (d)

Problem—93

One could expect the possibility of Reynolds number similarity in all of the following cases except

(a) submarines.
(b) torpedoes.
(c) seaplane hulls.
(d) supersonic aircraft.

The answer is (c)

Problem—94

One could expect the possibility of Reynolds number similarity in all of the following cases except

(a) pumps.
(b) fans.
(c) turbines.
(d) weirs.

The answer is (d)

Problem—95

One could expect the possibility of Froude number similarity in all of the following cases except

(a) surface ships.
(b) surface wave motion.
(c) flow over weirs.
(d) closed-pipe turbulent pipe flow.

The answer is (d)

Problem—96

One could expect the possibility of Froude number similarity in all of the following cases except

(a) motion of a fluid jet.
(b) flow over spillways.
(c) surge and flood waves.
(d) subsonic airfoils.

The answer is (d)

Chapter 6

Thermodynamics

Problem–1

A constant pressure process is

(a) isobaric.
(b) isothermal.
(c) adiabatic.
(d) isentropic.

The answer is (a)

Problem–2

A constant volume process is

(a) isobaric.
(b) isothermal.
(c) adiabatic.
(d) isometric.

The answer is (d)

Problem–3

A constant entropy process is

(a) isobaric.
(b) isothermal.
(c) isentropic.
(d) isometric.

The answer is (c)

Problem–4

A constant temperature process is

(a) isobaric.
(b) isothermal.
(c) adiabatic.
(d) isentropic.

The answer is (b)

Problem–5

A process with no heat transfer is

(a) isobaric.
(b) isothermal.
(c) adiabatic.
(d) isometric.

The answer is (c)

Problem–6

A thermodynamic process whose deviation from equilibrium is infinitesimal at all times is

(a) reversible.
(b) isentropic.
(c) in quasiequilibrium.
(d) isenthalpic.

The answer is (c)

Problem–7

Which thermodynamic property best describes the molecular activity of a substance?

(a) enthalpy
(b) entropy
(c) internal energy
(d) external energy

The answer is (c)

Problem–8

The combination of conditions that best describes a thermodynamic process is given by which of the following?

(I) has successive states through which the system passes

(II) when reversed leaves no change in the system

(III) when reversed leaves no change in the system or the surroundings

(IV) states are passed through so quickly that the surroundings do not change

 (a) I and II
 (b) I and III
 (c) I and IV
 (d) I only

The answer is (d)

Problem–9

The combination of conditions that best describes a reversible thermodynamic process is given by which of the following?

(I) has successive states through which the system passes

(II) when reversed leaves no change in the system

(III) when reversed leaves no change in the system or the surroundings

(IV) states are passed through so quickly that the surroundings do not change

 (a) I and II
 (b) I and III
 (c) I and IV
 (d) I only

The answer is (d)

Problem–10

A thermodynamic cycle can best be described by which combination of the following conditions?

(I) has successive states through which the system passes

(II) has successive states such that intial and final conditions are identical

(III) when reversed leaves no change in the surroundings

(IV) when reversed leaves no change in the system or the surroundings

 (a) I and II
 (b) I and III
 (c) I and IV
 (d) II and IV

The answer is (d)

Problem–11

A substance whose properties are uniform throughout is called a(n)

 (a) ideal substance.
 (b) solid.
 (c) standard substance.
 (d) pure substance.

The answer is (d)

Problem–12

All of the following are thermodynamic properties except

 (a) temperature.
 (b) pressure.
 (c) density.
 (d) modulus of elasticity.

The answer is (d)

Problem-13

A process that is adiabatic and reversible is also

(a) isobaric.
(b) isothermal.
(c) isentropic.
(d) isometric.

The answer is (c)

Problem-14

The volume of an ideal gas is halved, while its temperature is doubled. What happens to the pressure?

(a) The pressure doubles.
(b) The pressure is halved.
(c) The pressure is unchanged.
(d) The pressure is multiplied by 4.

The answer is (d)

Problem-15

A liquid boils when its vapor pressure equals

(a) the gage pressure.
(b) the critical pressure.
(c) the ambient pressure.
(d) one standard atmosphere.

The answer is (c)

Problem-16

A system composed of ice and water at 0°C is said to be

(a) a multiphase material.
(b) in thermodynamic equilibrium.
(c) in thermal equilibrium.
(d) all of the above

The answer is (d)

Problem-17

Let Q = heat entering a system
S = entropy
T = temperature
W = work
P = pressure
U = internal energy
V = volume

All of the following relations are true except

(a) $H = U + PV$
(b) $Q = \Delta U + W$
(c) $W = \int P dV$
(d) $S = \int Q dT$

The answer is (d)

Problem-18

The heat of fusion for a pure substance is

(a) the change in phase from solid to gas.
(b) the change in phase from liquid to gas.
(c) the energy released in a chemical reaction.
(d) the energy required to melt the substance.

The answer is (d)

Problem-19

The heat of vaporization involves the change in enthalpy due to

(a) the change in phase from solid to gas.
(b) the change in phase from liquid to gas.
(c) the energy released in a chemical reaction.
(d) the change in phase from solid to liquid.

The answer is (b)

Problem–20

The heat of sublimation involves the change in enthalpy due to

   (a) the change in phase from solid to gas.
   (b) the change in phase from liquid to gas.
   (c) the energy released in a chemical reaction.
   (d) the change in phase from solid to liquid.

The answer is (a)

Problem–21

A specific property

   (a) defines a specific variable (e.g., temperature).
   (b) is independent of mass.
   (c) is an extensive property multiplied by mass.
   (d) is dependent on the phase of the substance.

The answer is (b)

Problem–22

A material's specific heat can be defined as

   (a) the ratio of heat, Q, required to change the temperature of a mass, m, by a temperature, ΔT.
   (b) being different for constant pressure and constant temperature processes.
   (c) a function of temperature.
   (d) all of the above

The answer is (d)

Problem–23

If a substance's temperature is less than its saturation temperature, the substance is a

   (a) subcooled liquid.
   (b) liquid and a vapor.
   (c) saturated vapor.
   (d) superheated vapor.

The answer is (a)

Problem–24

If a substance's temperature is equal to its saturation temperature, the substance is a

   (a) subcooled liquid.
   (b) liquid and a vapor.
   (c) saturated vapor.
   (d) superheated vapor.

The answer is (b)

Problem–25

If a substance's temperature is greater than its saturation temperature, the substance is a

   (a) subcooled liquid.
   (b) liquid and a vapor.
   (c) saturated vapor.
   (d) superheated vapor.

The answer is (d)

Problem–26

Critical properties refer to

   (a) extremely important properties, such as temperature and pressure.
   (b) heats required for phase change and important for energy production.
   (c) property values where liquid and gas phases are indistinguishable.
   (d) properties having to do with equilibrium conditions, such as the Gibbs and Helmholtz functions.

The answer is (c)

Problem–27

For a saturated vapor, the relationship between temperature and pressure is given by

   (a) the perfect gas law.
   (b) van der Waal's equation.
   (c) the steam table.
   (d) a virial equation of state.

The answer is (c)

Problem—28

Properties of a superheated vapor are given by

   (a) the perfect gas law.

   (b) a superheat table.

   (c) a one-to-one relationship, such as the properties of saturated steam.

   (d) a virial equation of state.

The answer is (b)

Problem—29

Properties of nonreacting gas mixtures are given by

   (a) geometric weighting.

   (b) volumetric weighting.

   (c) volumetric weighting for molecular weight and density, and geometric weighting for all other properties except entropy.

   (d) arithmetic average.

The answer is (c)

Problem—30

The relationship between the total volume of a mixture of nonreacting gases and their partial volumes is given by

   (a) gravimetric fractions.

   (b) Amagat's law.

   (c) Dalton's law.

   (d) mole fractions.

The answer is (b)

Problem—31

The relationship between the total pressure of a mixture of nonreacting gases and the partial pressures of its constituents is given by

   (a) gravimetric fractions.

   (b) volumetric fractions.

   (c) Dalton's law.

   (d) mole fractions.

The answer is (c)

Problem—32

Which of the following is the best definition of enthalpy?

   (a) the ratio of heat added to the temperature increase in a system

   (b) the amount of useful energy in a system

   (c) the amount of energy no longer available to the system

   (d) the heat required to cause a complete conversion between two phases at a constant temperature

The answer is (b)

Problem—33

Which of the following statements is not true for real gases?

   (a) Molecules occupy a volume not negligible in comparison to the total volume of gas.

   (b) Real gases are subject to attractive forces between molecules (e.g., van der Waal's forces).

   (c) The law of corresponding states may be used for real gases.

   (d) Real gases are found only rarely in nature.

The answer is (d)

Problem—34

All of the following processes ($\leftrightarrow$ means one implies the other) are correct except

   (a) throttling $\leftrightarrow$ constant enthalpy.

   (b) isentropic $\leftrightarrow$ constant temperature.

   (c) isobaric $\leftrightarrow$ constant pressure.

   (d) adiabatic $\leftrightarrow$ isentropic.

The answer is (b)

Problem–35

For a polytropic process in which $p_1(V_1)^n = p_2(V_2)^n$, all of the values of n given below are correct except

(a) $n = 0$, isobaric
(b) $n = 1$, isothermal
(c) $n = k$, isentropic
(d) $n = \dfrac{k-1}{k}$, isenthalpic

The answer is (d)

Problem–36

All of the following processes are irreversible except

(a) stirring of a viscous fluid.
(b) a isentropic deceleration of a moving, perfect fluid.
(c) an unrestrained expansion of a gas.
(d) phase changes.

The answer is (b)

Problem–37

All of the following processes are irreversible except

(a) chemical reactions.
(b) diffusion.
(c) current flow through an electrical resistance.
(d) an isentropic compression of a perfect gas.

The answer is (d)

Problem–38

All of the following processes are irreversible except

(a) magnetization with hysteresis.
(b) elastic tension and release of a steel bar.
(c) inelastic deformation.
(d) heat conduction.

The answer is (b)

Problem–39

Each of the following is a point function (values independent of the path) except

(a) p, pressure.
(b) T, temperature.
(c) W, work.
(d) s, entropy.

The answer is (c)

Problem–40

Which of the following state(s) is (are) necessary for a system to be in thermodynamic equilibrium?

(a) chemical equilibrium
(b) thermal equilibrium
(c) mechanical equilibrium
(d) chemical, mechanical, and thermal equilibrium

The answer is (d)

Problem–41

The first law of thermodynamics for a closed system is $Q = \Delta U + W$. The sign convention is

(a) Q positive in, W positive in, and ΔU positive for increased internal energy.
(b) Q positive out, W positive in, and ΔU negative for decreased internal energy.
(c) Q positive out, W positive out, and ΔU negative for increased internal energy.
(d) Q positive in, W positive out, and ΔU negative for decreased internal energy.

The answer is (d)

Problem–42

The heat transfer term in the first law of thermodynamics may be due to any of the following except

(a) conduction.
(b) convection.
(c) radiation.
(d) internal heat generation (e.g., chemical reaction).

The answer is (d)

Problem–43

A system that experiences no mass crossing the system boundaries (in or out) is most properly called a(n)

(a) open system.
(b) closed system.
(c) quasisteady system.
(d) ideal system.

The answer is (b)

Problem–44

A system in which a substance is allowed to enter and leave is most properly called a(n)

(a) open system.
(b) closed system.
(c) quasisteady system.
(d) unsteady (dynamic) system.

The answer is (a)

Problem–45

The first and second laws of thermodynamics are

(a) continuity equations.
(b) momentum equations.
(c) energy equations.
(d) equations of state.

The answer is (c)

Problem–46

A constant pressure thermodynamic process obeys

(a) Boyles' law.
(b) Charles' law.
(c) Amagat's law.
(d) Dalton's law.

The answer is (b)

Problem–47

A constant temperature thermodynamic process obeys

(a) Boyles' law.
(b) Charles' law.
(c) Amagat's law.
(d) Dalton's law.

The answer is (a)

Problem–48

Which of the following relationships is untrue?

(a) constant volume process of an ideal gas

$$p_2 = p_1 \left(\frac{T_2}{T_1} \right)$$

(b) constant pressure process of an ideal gas

$$T_2 = T_1 \left(\frac{V_2}{V_1} \right)$$

(c) constant temperature process of an ideal gas

$$V_2 = V_1 \left(\frac{p_1}{p_2} \right)$$

(d) constant enthalpy process of an ideal gas

$$p_2 > p_1$$

The answer is (d)

Problem–49

A series of processes that eventually bring the system back to its original condition is called a

(a) reversible process.
(b) irreversible process.
(c) cycle.
(d) isentropic process.

The answer is (c)

Problem–50

An inventor proposes to develop electrical power by withdrawing heat from the geyser fields of northern California and converting it all to work in power turbines. This scheme will not work because

(a) the geyser fields have only a limited lifetime.
(b) the salinity of the steam is too great.
(c) it violates the first law of thermodynamics.
(d) it violates the second law of thermodynamics.

The answer is (d)

Problem–51

A process that is adiabatic and reversible is also

(a) isometric.
(b) isobaric.
(c) isentropic.
(d) isothermal.

The answer is (c)

Problem–52

Which of the following statements about a path function is not true?

(a) On a p-V diagram, it can represent work done.
(b) On a T-s diagram, it can represent heat transferred.
(c) It is dependent on the path between states of thermodynamic equilibrium.
(d) It represents values of p, V, T, and s between states that are path functions.

The answer is (d)

Problem–53

The following terms are included in the first law of thermodynamics for closed systems except

(a) heat transferred in and out of the system.
(b) work done by or on the system.
(c) internal energy.
(d) kinetic energy.

The answer is (d)

Problem–54

The following terms are included in the first law of thermodynamics for open systems except

(a) heat transferred in and out of the system.
(b) work done by or on the system.
(c) magnetic energy.
(d) internal energy.

The answer is (c)

Problem–55

All of the following terms are included in the second law for open systems except

(a) shaft work—work a steady flow device does on the surroundings.
(b) flow work—p-V work.
(c) internal energy.
(d) mixing work.

The answer is (d)

Problem–56

All of the following are Maxwell's relations except

(a) $\left(\dfrac{\partial T}{\partial V}\right)_s = -\left(\dfrac{\partial p}{\partial s}\right)_V$

(b) $\left(\dfrac{\partial T}{\partial p}\right)_s = \left(\dfrac{\partial V}{\partial s}\right)_p$

(c) $\left(\dfrac{\partial s}{\partial V}\right)_T = \left(\dfrac{\partial p}{\partial T}\right)_V$

(d) $\left(\dfrac{\partial M}{\partial y}\right)_x = \left(\dfrac{\partial N}{\partial x}\right)_y$

The answer is (d)

Problem-57

Which of the following is not a correct statement about the graphical representation of work and heat?

(a) The values of p, V, T, and s are point functions because their values are independent of the path taken to the thermodynamic state.

(b) Work and heat are also point functions because their values are independent of the path taken to the thermodynamic state.

(c) The work done by or on the system can be found by integrating the area under the path of the p-V curve.

(d) The amount of heat absorbed or released from a system can be found by integrating the area under the path on a T-s diagram.

The answer is (b)

Problem-58

Which of the following is not a correct statement of the laws of thermodynamics?

(a) During a process, the net heat transfer out of a system is equal to the net work output less the change in energy.

(b) It is impossible to construct a refrigerator that, operating in a cycle, will produce no effect other than the transfer of heat from a cooler to a hotter body.

(c) A natural process that starts in one equilibrium state and ends in another will go in the direction that increases the entropy of the system and the environment.

(d) It is impossible to operate an engine in a cycle that will have no other effect than to extract heat from a reservoir and turn it into an equivalent amount of work.

The answer is (a)

Problem-59

The following factors are necessary to define a thermodynamic cycle except

(a) the working substance.
(b) high and low temperature reservoirs.
(c) the time it takes to complete the cycle.
(d) the means of doing work on the system.

The answer is (c)

Problem-60

The maximum possible work that can be obtained from a cycle operating between two reservoirs is found from

(a) process irreversibility.
(b) availability.
(c) Carnot efficiency.
(d) reversible work.

The answer is (c)

Problem-61

All of the following mechanisms can supply heat to a thermodynamic system except

(a) conduction.
(b) natural convection.
(c) adiabatic expansion.
(d) radiation.

The answer is (c)

Problem-62

All heat transfer processes require a medium of energy exchange except

(a) conduction.
(b) natural convection.
(c) forced convection.
(d) radiation.

The answer is (d)

Problem–63

Thermal conduction is described by

   (a) Newton's law.
   (b) the logarithmic mean temperature difference.
   (c) the Stefan-Boltzmann law.
   (d) Fourier's law.

The answer is (d)

Problem–64

Convection is described by

   (a) Newton's law.
   (b) the logarithmic mean temperature difference.
   (c) the Stefan-Boltzmann law.
   (d) Fourier's law.

The answer is (a)

Problem–65

Radiant heat transfer is described by

   (a) Newton's law.
   (b) the Stefan-Boltzmann law.
   (c) Fourier's law.
   (d) Kirchhoff's law.

The answer is (b)

Problem–66

The equivalence of ratios of emissive power to absorbtivity for bodies in thermal equilibrium is described by

   (a) Newton's law.
   (b) the Stefan-Boltzmann law.
   (c) Fourier's law.
   (d) Kirchhoff's law.

The answer is (d)

Problem–67

The temperature potential between temperatures at the two ends of a heat exchanger are given by

   (a) the logarithmic mean temperature difference.
   (b) the Stefan-Boltzmann law.
   (c) Fourier's law.
   (d) Kirchhoff's law.

The answer is (a)

Problem–68

The determination of properties and behavior of atmospheric air is usually the purview of

   (a) thermodynamics.
   (b) psychrometrics.
   (c) forced convection.
   (d) Kirchhoff's laws.

The answer is (b)

Problem–69

All of the following temperatures have meaning in psychrometrics except

   (a) dry-bulb temperature.
   (b) wet-bulb temperature.
   (c) adiabatic wall temperature.
   (d) dew point.

The answer is (c)

Refer to the following illustration for Probs. 70 through 74.

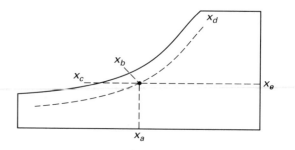

Problem–70

The dry-bulb temperature is given by

   (a) x_a.
   (b) x_b.
   (c) x_c.
   (d) x_d.

The answer is (a)

Problem–71

The wet-bulb temperature is given by

   (a) x_a.
   (b) x_b.
   (c) x_c.
   (d) x_d.

The answer is (b)

Problem–72

The relative humidity is given by

   (a) x_a.
   (b) x_b.
   (c) x_c.
   (d) x_d.

The answer is (d)

Problem–73

The specific humidity is given by

   (a) x_a.
   (b) x_b.
   (c) x_c.
   (d) x_e.

The answer is (d)

Problem–74

The dew point is given by

   (a) x_a.
   (b) x_b.
   (c) x_c.
   (d) x_e.

The answer is (c)

Problem–75

The relative humidity is given by the

   (a) ratio of the actual humidity to the saturated humidity at the same temperature and pressure.
   (b) ratio of the partial pressure of water vapor to the saturation pressure.
   (c) ratio of wet-bulb to dry-bulb temperature.
   (d) ratio of dry-bulb temperature to dew point.

The answer is (b)

Problem–76

All of the following processes can be found on a psychrometric chart except

   (a) humidification.
   (b) sensible heating.
   (c) natural convection.
   (d) sensible cooling.

The answer is (c)

Problem–77

All of the following processes can be found on a psychrometric chart except

   (a) heating and humidifying.
   (b) cooling and dehumidification.
   (c) black body radiation.
   (d) evaporative cooling.

The answer is (d)

Problem–78

The function of a pump or compressor is to

(a) transfer heat from one fluid to another.
(b) increase the total energy content of the flow.
(c) extract energy from the flow.
(d) exchange heat to increase energy to the flow.

The answer is (b)

Problem–79

The function of a heat exchanger is to

(a) increase the water temperature entering the boiler and decrease combustion requirements.
(b) transfer heat from one fluid to another.
(c) increase the total energy content of the flow.
(d) exchange heat to increase energy to the flow.

The answer is (b)

Problem–80

The function of a preheater is to

(a) increase the water temperature entering the boiler and decrease combustion requirements.
(b) transfer heat from one fluid to another.
(c) increase the total energy content of the flow.
(d) exchange heat to increase energy to the flow.

The answer is (a)

Problem–81

The function of a turbine is to

(a) transfer heat from one fluid to another.
(b) increase the total energy content of the flow.
(c) extract energy from the flow.
(d) exchange heat to increase energy to the flow.

The answer is (c)

Problem–82

The function of a superheater is to

(a) increase the water temperature entering the boiler and decrease combustion requirements.
(b) transfer heat from one fluid to another.
(c) increase the total energy content of the flow.
(d) exchange heat to increase energy to the flow.

The answer is (d)

Problem–83

Adiabatic heat transfer within a vapor cycle refers to

(a) heat transfer that is isentropic but not reversible.
(b) the transfer of energy from one stream to another in a heat exchanger so that the energy of the input streams equals the energy of the output streams.
(c) heat transfer that is reversible but not isentropic.
(d) There is no such thing as adiabatic heat transfer.

The answer is (b)

Refer to the following illustrations for Probs. 84 through 86.

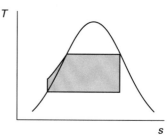

(a)

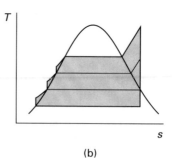

(b)

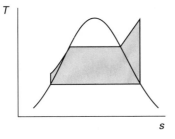

(c)

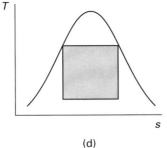

(d)

Problem–84

The basic Rankine cycle is shown by

   (a) a.
   (b) b.
   (c) c.
   (d) d.

The answer is (a)

Problem–85

The Rankine cycle with superheat is shown by

   (a) a.
   (b) b.
   (c) c.
   (d) d.

The answer is (c)

Problem–86

The Carnot cycle is shown by

   (a) a.
   (b) b.
   (c) c.
   (d) d.

The answer is (d)

Problem–87

What is the maximum efficiency of a power cycle operating between $727°C$ and $177°C$?

   (a) 35%
   (b) 45%
   (c) 55%
   (d) 68%

The answer is (c)

Refer to the following illustrations for Probs. 88 through 91.

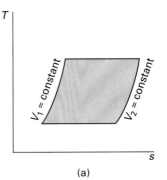

(a)

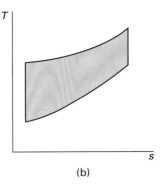

(b)

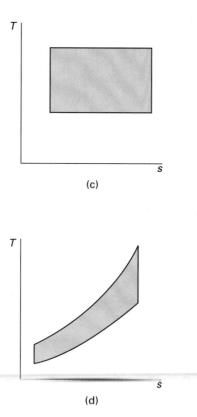

(c)

(d)

Problem–88

The air standard Carnot cycle is shown by

(a) a.
(b) b.
(c) c.
(d) d.

The answer is (c)

Problem–89

The air standard Otto cycle is shown by

(a) a.
(b) b.
(c) c.
(d) d.

The answer is (d)

Problem–90

The air standard diesel cycle is shown by

(a) a.
(b) b.
(c) c.
(d) d.

The answer is (b)

Problem–91

The air standard Stirling cycle is shown by

(a) a.
(b) b.
(c) c.
(d) d.

The answer is (a)

Problem–92

A Carnot engine receives 1000 Btu from a hot reservoir at 600°F and rejects 360 Btu of heat. What is the efficiency of the engine?

(a) 36%
(b) 50%
(c) 64%
(d) 80%

The answer is (c)

Problem–93

Which of the following statements about air standard cycles is not correct?

(a) They use a fixed amount of air as the working fluid.
(b) They are similar to vapor cycles because combustion products can be returned to their initial conditions for reuse.
(c) Their working fluid, air, is considered an ideal gas with constant specific heat.
(d) Their combustion process is replaced by a process of instantaneous heat transfer from high temperature surroundings.

The answer is (b)

Problem–94

The Carnot refrigeration cycle includes all of the following processes except

- (a) isentropic expansion.
- (b) isothermal heating.
- (c) isenthalpic expansion.
- (d) isentropic compression.

The answer is (c)

Problem–95

The vapor cycle is to thermal efficiency as the refrigeration cycle is to the

- (a) energy efficiency ratio.
- (b) COP for a refrigerator.
- (c) COP for a heat pump.
- (d) Carnot efficiency.

The answer is (b)

Problem–96

Which of the following expressions for the efficiency of a reversible thermodynamic cycle is not true?

(a) $\eta = \dfrac{W_{\text{out}} - W_{\text{in}}}{Q_{\text{in}}}$

(b) $\eta = \dfrac{Q_{\text{in}} - Q_{\text{out}}}{Q_{\text{in}}}$

(c) $\eta = \dfrac{Q_{\text{net}}}{Q_{\text{in}}}$

(d) $\eta = \dfrac{Q_{\text{out}} - Q_{\text{in}}}{Q_{\text{in}}}$

The answer is (d)

Problem–97

The isentropic efficiency of a turbine is given by

- (a) the ratio of actual to ideal energy extracted.
- (b) the ratio of actual to ideal energy inputted.
- (c) the ratio of ideal to actual energy extracted.
- (d) none of the above

The answer is (a)

Problem–98

The isentropic efficiency of a pump is given by the

- (a) ratio of actual to ideal energy extracted.
- (b) ratio of ideal to actual energy inputted.
- (c) ratio of ideal to actual energy extracted.
- (d) ratio of actual to ideal energy inputted.

The answer is (b)

Problem–99

The work required by a pump to increase water pressure from 1 MPa (1 atm) to 18 MPa (18 atm) is 20 kJ/kg. What is the efficiency of the pump?

- (a) 80%
- (b) 85%
- (c) 90%
- (d) 95%

The answer is (b)

Refer to the following illustrations for Probs. 100 through 103.

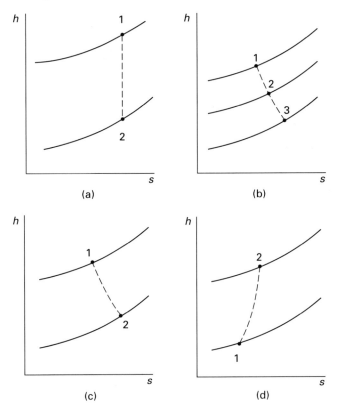

(a)   (b)

(c)   (d)

Problem—100

The actual compression of a pump is shown by

- (a) a.
- (b) b.
- (c) c.
- (d) d.

The answer is (d)

Problem—101

The ideal expansion across a turbine is shown by

- (a) a.
- (b) b.
- (c) c.
- (d) d.

The answer is (a)

Problem—102

The actual expansion across a turbine is shown by

- (a) a.
- (b) b.
- (c) c.
- (d) d.

The answer is (c)

Problem—103

The actual expansion across a two-stage turbine is shown by

- (a) a.
- (b) b.
- (c) c.
- (d) d.

The answer is (b)

Problem—104

A Mollier chart of thermodynamic properties is shown on a(n)

- (a) T-s diagram.
- (b) p-V diagram.
- (c) h-s diagram.
- (d) p-h diagram.

The answer is (c)

Problem—105

The electrical efficiency of a device is the ratio of the

- (a) electrical energy output to the electrical energy input.
- (b) mechanical energy input to the mechanical energy output of the device.
- (c) actual energy extracted to the ideal energy extracted.
- (d) actual to the ideal energy input.

The answer is (a)

Problem—106

The mechanical efficiency of a device is the ratio of the

- (a) mechanical energy input to the mechanical energy output of the device.
- (b) ideal energy input to the actual energy input.
- (c) actual energy extracted to the ideal energy extracted.
- (d) actual to the ideal energy input.

The answer is (b)

Problem–107

The adiabatic pump efficiency is the ratio of the

(a) electrical energy output to the electrical energy input.
(b) mechanical energy input to the mechanical energy output of the device.
(c) ideal energy input to the pump to the actual energy input.
(d) actual energy extracted to the ideal energy extracted.

The answer is (c)

Problem–108

The adiabatic turbine efficiency is the ratio of the

(a) electrical energy output to the electrical energy input.
(b) mechanical energy input to the mechanical energy output of the device.
(c) ideal energy input to the pump to the actual energy input.
(d) actual energy extracted to the ideal energy extracted.

The answer is (d)

Problem–109

During a thermal process, 30 J of work is done by a closed system on its surroundings. The internal energy of the system decreases by 50 J. What is the net heat transfer?

(a) 20 J released into the surroundings
(b) 20 J absorbed into the system
(c) 80 J released into the surroundings
(d) 80 J absorbed into the system

The answer is (a)

Problem–110

The compression ratio of an air standard Otto cycle is the ratio of

(a) pressures.
(b) cylinder volumes.
(c) actual gas compression to ideal gas compression.
(d) temperatures in an isentropic compression.

The answer is (b)

Problem–111

The compression ratio in a gas compression process is the ratio of

(a) pressures.
(b) cylinder volumes.
(c) actual gas compression to ideal gas compression.
(d) none of the above

The answer is (a)

Problem–112

The basis for an intercooler is that

(a) for a given compression ratio, work for an isothermal compression is less than work for an adiabatic compression.
(b) for a given compression ratio, work for an isothermal compression is greater than work for an adiabatic compression.
(c) it is easier to remove heat after gas has been compressed than before.
(d) intercoolers are cheap to manufacture.

The answer is (a)

Problem–113

What is the relationship between the COP of a refrigerator and a heat pump?

(a) $(COP)_{refrigerator} = (COP)_{heat\ pump}$

(b) $(COP)_{refrigerator} = \dfrac{1}{(COP)_{heat\ pump}}$

(c) $(COP)_{refrigerator} = 1 - \dfrac{1}{(COP)_{heat\ pump}}$

(d) $(COP)_{refrigerator} = (COP)_{heat\ pump} - 1$

The answer is (d)

Problem–114

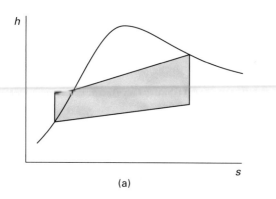

(a)

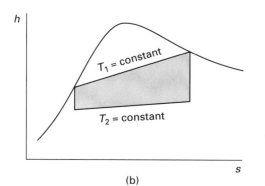

(b)

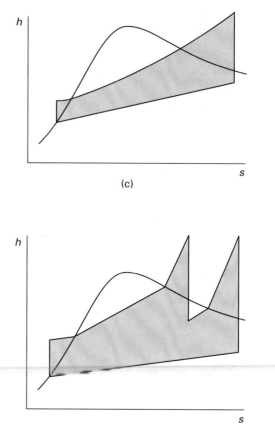

(c)

(d)

All of the cycles shown represent a basic Rankine cycle except

(a) a.
(b) b.
(c) c.
(d) d.

The answer is (b)

Chapter 7

Chemistry

Problem–1

The smallest subdivision of an element that can take place in a chemical reaction is a(n)

✓(a) atom.
(b) molecule.
(c) electron.
(d) proton.

The answer is (a)

Problem–2

The smallest subdivision of a compound that can exist in a natural state is a(n)

(a) atom.
✓(b) molecule.
(c) electron.
(d) proton.

The answer is (b)

Problem–3

Elements with different atomic weights but the same atomic number are

(a) isomers.
(b) isotropes.
✓(c) isotopes.
(d) isobars.

The answer is (c)

Problem–4

All of the following are characteristics of metals except

(a) high electrical conductivities
(b) tendency to form positive ions
✓(c) tendency to form brittle solids
(d) high melting points

The answer is (c)

Problem–5

The following are all characteristics of nonmetals except

(a) having little or no luster
(b) appearing on the right end of the periodic table
(c) having low ductility
(d) being reducing agents

The answer is (d)

Problem–6

The tendency of a pure compound to be composed of the same elements combined in a definite proportion by mass is

(a) Avogadro's law.
(b) Boyles law.
(c) the law of definite proportions.
(d) Le Châtelier's principle.

The answer is (c)

Problem–7

In an oxidation-reduction chemical reaction, all of the following occur except

   (a) the exchange of electrons between elements.
   (b) elements becoming more positive.
   (c) elements becoming more negative.
   ✓(d) nuclear fusion.

The answer is (d)

Problem–8

Which of the following statements involving chemical reactions is not true?

   (a) A compound can be a liquid, gas, or solid.
   (b) Ions can be positive or negative.
   (c) A compound can be formed from elements by chemical means.
   ✓(d) The same element can have different atomic numbers.

The answer is (d)

Problem–9

Atomic weights of the elements in the periodic table are not whole numbers because of

   (a) the existence of isotopes.
   (b) imprecise measurements during the development of the periodic table.
   (c) round-off error in calculating atomic weights.
   (d) the change of reference of the atomic mass unit from oxygen-16 to carbon-12 in 1961.

✓ The answer is (a)

Problem–10

Graduations in the properties of elements from one element to the next are less pronounced in

   (a) the lanthanide series.
   (b) periods.
   ✓(c) groups.
   (d) active metals.

The answer is (c)

Problem–11

What is the ground state of the electron configuration of titanium (Ti, $Z = 22$)?

   (a) $1s^2 2s^2 2p^6 3s^2 3p^6 4s^2 4p^2$
   (b) $1s^2 2s^2 2p^6 3s^2 3p^4 4s^2 4p^4$
   (c) $1s^2 2s^2 2p^6 3s^2 3p^4 4d^6$
   ✓(d) $1s^2 2s^2 2p^6 3s^2 3p^6 4s^2 3d^2$

The answer is (d)

Problem–12

The valences in the following table are all correct except

	Group	Name	Valence
(a)	IA	alkali metals	$+1$
(b)	VIIA	halogens	-1
(c)	O	noble gases	0
✓(d)	VA	metalloids	$+3$

The answer is (d)

Problem–13

All of the following are components of a chemical element except

   (a) protons.
   (b) neutrons.
   (c) electrons.
   ✓(d) ions.

The answer is (d)

Problem–14

Which one of the following lists contains only metals.?

   (a) magnesium (Mg), copper (Cu), selenium (Se)
   (b) carbon (C), tellurium (Te), tantalum (Ta)
   ✓(c) beryllium (Be), magnesium (Mg), cobalt (Co)
   (d) sulfur (S), lead (Pb), rubidium (Rb)

The answer is (c)

Problem–15

The following properties are different for isomers of the same chemical compound except

(a) density.
(b) melting point.
(c) number of atoms in a mole of each isomer.
(d) specific heat.

The answer is (c)

Problem–16

Which of the following is not a prefix used in naming isomers?

(a) para-
(b) meta-
(c) cis-
(d) bi-

The answer is (d)

Problem–17

The two molecules shown are which of the following?

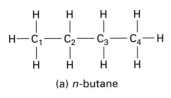

(a) n-butane

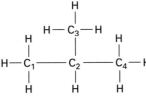

(b) isobutane

(a) isotopes
(b) atoms
(c) isomers
(d) atomic particles

The answer is (c)

Problem–18

All of the following are types of chemical bonds except

(a) ionic bonds.
(b) covalent bonds.
(c) metallic bonds.
(d) nuclear bonds.

The answer is (d)

Problem–19

The equilibrium distance between elements in an ionic bond is a function of all the following except

(a) ionic charge.
(b) coordination number.
(c) atomic weight.
(d) temperature.

The answer is (c)

Problem–20

Which of the following statements is not a characteristic of ionic compounds?

(a) They are usually hard, brittle, crystalline solids.
(b) They have high melting points.
(c) They are nonvolatile and have low vapor pressures.
(d) They are good electrical conductors in the solid phase.

The answer is (d)

Problem–21

Which of the following statements is not a characteristic of ionic compounds?

(a) They become electrically conductive when dissociated into the component ions.
(b) Those that are soluble in water form electrolytic solutions that conduct electricity (i.e., they are electrolytes).
(c) They have small lattice energies due to weak electrostatic attractive force.
(d) They have high melting points.

The answer is (c)

Problem–22

What kind of bonding do common gases that exist in a free state as diatomic molecules experience?

    (a) ionic bonds

√ (b) covalent bonds

    (c) metallic bonds

    (d) nuclear bonds

The answer is (b)

Problem–23

When electrons are not shared equally between two elements, and electrons spend more time with one element than with the other, the bonding is called

    (a) ionic bonding.

    (b) polar covalent bonding.

    (c) nonpolar covalent bonding.

    (d) metallic bonding.

The answer is (b)

Problem–24

When electrons are shared equally (e.g., when the atoms are the same as in diatomic gases), the bonding is called

    (a) ionic bonding.

    (b) polar covalent bonding.

    (c) nonpolar covalent bonding.

    (d) resonance bonding.

The answer is (c)

Problem–25

All of the following are common types of chemical reactions except

    (a) direct combination.

    (b) fission.

    (c) decomposition.

    (d) double replacement.

The answer is (b)

Problem–26

Redox reactions can often be a type of

    (a) direct combination.

    (b) fission.

    (c) decomposition.

    (d) double replacement.

The answer is (d)

Problem–27

All of the following occur during oxidation of a substance except

    (a) oxidation state increases.

    (b) loss of electrons.

    (c) the substance becomes less negative.

    (d) oxidation of the oxidizing agent.

The answer is (d)

Problem–28

All of the following occur during reduction of a substance except

    (a) an increase in negative charge.

    (b) loss of electrons.

    (c) an oxidation state decrease.

    (d) reduction of the oxidizing agent.

The answer is (b)

Problem–29

Which of the following does not occur during a stoichiometric reaction?

    (a) The number of electrons lost by some elements equals the number of electrons gained by other elements.

    (b) The number of elements on one side of the equation equals the number of elements on the other.

    (c) There are either more reactants than products or more products than reactants so that some compounds do not participate in the reaction at all.

    (d) Solutions to weight and proportion problems can be solved by simple ratios.

The answer is (c)

Problem–30

Which of the following statements is not true of stoichiometric reactions?

(a) Just the right amount of reactants are available.
(b) When the reaction goes to completion, there are no unused reactants.
(c) The number of moles of elements in the reactants equals the number of moles in the products.
(d) The reactant that is used up first is called the limiting reactant.

The answer is (d)

Problem–31

All of the following are true of nonstoichiometric reactions except

(a) there is an excess of one or more reactants.
(b) the percentage yield measures the efficiency of the reaction.
(c) nonstoichiometric reactions are rare in the combustion process.
(d) in combustion, air is often the excess reactant to assure complete combustion of fuel.

The answer is (c)

Problem–32

The fact that the amount of slightly soluble gas absorbed in a liquid is proportional to the partial pressure of the gas is known as

(a) Dalton's law.
(b) Henry's law.
(c) Raoult's law.
(d) Boyle's law.

The answer is (b)

Problem–33

If the solute particles of a solid suspended in a liquid are larger than molecules, the mixture is known as a

(a) solution.
(b) suspension.
(c) hydration.
(d) saturated solution.

The answer is (b)

Problem–34

When a solvent has dissolved as much solute as it can, the mixture is called a

(a) solution.
(b) suspension.
(c) hydration.
(d) saturated solution.

The answer is (d)

Problem–35

When excess solute in a solution settles to the bottom of the container, the process is called

(a) solvation.
(b) deemulsification.
(c) precipitation.
(d) aquation.

The answer is (c)

Problem–36

All of the following express units of concentration except

(a) normality.
(b) molarity.
(c) formality.
(d) isotropy.

The answer is (d)

Problem–37

The amount of energy absorbed when a substance enters a solution is called the

(a) heat of fusion.
(b) heat of sublimation.
(c) endothermic heat of solution.
(d) exothermic heat of solution.

The answer is (c)

Problem–38

The absorption of water by a dessicant often demonstrates

(a) heat of fusion.
(b) heat of vaporization.
(c) endothermic heat of solution.
(d) exothermic heat of solution.

The answer is (d)

Problem–39

A substance that absorbs moisture from the air is

(a) deliquescent.
(b) efflorescent.
(c) effervescent.
(d) a precipitant.

The answer is (a)

Problem–40

The drop in a solvent's vapor pressure, and the increase in mole fraction as solute is added is described by

(a) Dalton's law.
(b) Henry's law.
(c) Raoult's law.
(d) Boyle's law.

The answer is (c)

Problem–41

All of the following statements are characteristics of bases except

(a) they conduct electricity in aqueous solutions.
(b) they turn red litmus paper blue.
(c) they have a pH between 0 and 7.
(d) they neutralize acids forming salts and water.

The answer is (c)

Problem–42

All of the following statements are characteristics of acids except

(a) they do not conduct electricity in aqueous solutions.
(b) they turn blue litmus paper red.
(c) they have a pH between 0 and 7.
(d) they neutralize bases forming salts and water.

The answer is (a)

Problem–43

All of the following statements about conjugate acids and bases are true except

(a) a conjugate acid results when a base accepts a proton.
(b) a conjugate base results when a base accepts a proton.
(c) strong acids tend to give weak conjugate bases.
(d) the Brönsted-Lowry theory defines bases as proton acceptors.

The answer is (b)

Problem–44

All of the following factors affect rates of reaction except

(a) exposed surface area.
(b) concentrations.
(c) temperature.
(d) pressure.

The answer is (d)

Problem–45

Le Châtelier's principle predicts the direction of a state of chemical equilibrium based on all of the following factors except

(a) temperature.
(b) specific volume.
(c) pressure.
(d) concentration.

The answer is (b)

Problem–46

The amount of energy necessary to cause a reaction to occur is called the

(a) heat of formation.
(b) heat of solution.
(c) activation energy.
(d) heat of fusion.

The answer is (c)

Problem–47

The effect of a catalyst in a chemical reaction is to

(a) absorb the exothermic heat of reaction.
(b) provide the exothermic heat of reaction.
(c) lower the activation energy.
(d) provide the heat of sublimation.

The answer is (c)

Problem–48

The relationship between the concentrations of products and reactants in a reversible chemical reaction is given by

(a) the ionization constant.
(b) the equilibrium constant.
(c) the solubility product.
(d) Le Châtelier's principle.

The answer is (b)

Problem–49

The equilibrium constant for weak solutions is known as

(a) the ionization constant.
(b) the Arrhenius exponent.
(c) the solubility product.
(d) Le Châtelier's principle.

The answer is (a)

Problem–50

The speed at which a reaction proceeds to equilibrium is the purview of

(a) reaction kinetics.
(b) Le Châtelier's principle.
(c) neutralization.
(d) ionization.

The answer is (a)

Problem–51

All of the following are units of energy except

(a) atomic units.
(b) MeV.
(c) dynes.
(d) ergs.

The answer is (c)

Problem–52

All of the following are units of energy except

(a) calories.
(b) joules.
(c) pascals.
(d) MeV.

The answer is (c)

Problem–53

The first person credited with recognizing the structure of electromagnetic waves was

   (a) Max Planck in 1864.
   (b) Albert Einstein in 1894.
   (c) Marie Curie in 1882.
   (d) James Clerk Maxwell in 1864.

  The answer is (d)

Problem–54

Most observed properties of light and other radiant energy are consistent with waves in nature, but in interactions with matter, electromagnetic energy behaves as though it consists of discrete pieces or

   (a) blocks.
   (b) balls.
   (c) quanta.
   (d) atomic masses.

  The answer is (c)

Problem–55

All of the following terms are synonymous with quanta of electromagnetic theory except

   (a) packets.
   (b) corpuscles.
   (c) x-rays.
   (d) photons.

  The answer is (c)

Problem–56

The relationship between the frequency f, the wavelength λ, and the speed of light c for electromagnetic waves is given by which of the following?

   (a) $c = \dfrac{\lambda}{f}$

   (b) $c = \dfrac{f}{\lambda}$

   (c) $c = f\lambda$

   (d) $\dfrac{cf}{\lambda} = 1$

  The answer is (c)

Problem–57

The colors of the visible spectrum, ranging from shortest to longest wavelength, are given by

   (a) red, yellow, orange, blue, green, and violet.
   (b) yellow, orange, green, violet, red, and blue.
   (c) violet, blue, green, yellow, orange, and red.
   (d) red, yellow, orange, green, violet, and blue.

  The answer is (c)

Problem–58

The relationship between a quantum of energy E and the frequency of electromagnetic radiation f is given by $E = hf$ where h is

   (a) system enthalpy.
   (b) hybrid parameter.
   (c) Planck's constant.
   (d) angular momentum.

  The answer is (c)

Problem–59

The phenomenon in which a short wavelength photon (gamma-ray, x-ray, etc.) hits an atom on the surface of a substance and causes an electron to be ejected is called the

   (a) photoelectric effect.
   (b) Seebeck effect.
   (c) greenhouse effect.
   (d) Peltier effect.

  The answer is (a)

Problem–60

Einstein reasoned there was a discrete amount of energy needed to remove an electron from a surface, with the rest of the incident photon's energy contributing to the kinetic energy of the photon. The amount of energy is called the

   (a) binding energy.
   (b) work function.
   (c) Coulomb energy.
   (d) Fermi energy.

  The answer is (b)

Problem–61

If the energy of the incident photon is less than the work function,

    (a) an electron will be ejected.
    (b) more than one electron will be ejected.
    (c) an electron will not be ejected.
    (d) less than one electron will be ejected.

The answer is (c)

Problem–62

Rutherford's experiment in 1911 with collimated alpha particles from radium and gold foil helped him calculate

    (a) the charge of the alpha particle.
    (b) the mass of the alpha particle.
    (c) the electrostatic radius of the gold nucleus.
    (d) the kinetic energy of the alpha particle.

The answer is (c)

Problem–63

The number of molecules per unit volume of a material is given in terms of the material density ρ, Avogadro's number N_A, and the material's atomic weight AW as which of the following?

    (a) $\rho N_A(\text{AW})$
    (b) $\dfrac{\rho N_A}{\text{AW}}$
    (c) $\dfrac{\text{AW}}{\rho N_A}$
    (d) $\dfrac{N_A}{\rho(\text{AW})}$

The answer is (b)

Problem–64

The time required for half a quantity of radioactive particles to decay (disintegrate) is called its

    (a) average life.
    (b) median life.
    (c) time constant.
    (d) half life.

The answer is (d)

Problem–65

Which of the following is not a postulate of Bohr's theory of the hydrogen atom?

    (a) Electron orbits are discrete and nonradiating, and an electron may not remain between these orbits.
    (b) The energy change experienced by an electron changing from one orbit to another is quantized.
    (c) Light waves exist simultaneously as high-frequency electrical and magnetic waves.
    (d) Angular momentum is quantized.

The answer is (c)

Problem–66

The statement that the product of the error in the measured determination of a particle's position and its momentum is of the order of Planck's constant h is known as

    (a) Bohr's theory.
    (b) D'Alembert's paradox.
    (c) the Heisenberg uncertainty principle.
    (d) Planck's law.

The answer is (c)

Problem–67

The mathematical description that overcomes the deficiencies of the Bohr atom is the

    (a) Heisenberg principle.
    (b) Schrödinger wave equation.
    (c) Pauli exclusion principle.
    (d) Bohr reconsideration principle.

The answer is (b)

Problem–68

At present, the number of true elementary particles, which include leptons and quarks, is thought to be

    (a) 4.
    (b) 8.
    (c) 10.
    (d) 12.

The answer is (d)

Problem–69

The effective size of a target atom that interacts with a moving particle is called its

    (a) length.
    (b) width.
    (c) cross section.
    (d) pseudo-area.

The answer is (c)

Problem–70

Most nuclear particles can react with atoms in several different ways including

    (a) absorption.
    (b) scattering.
    (c) absorption and scattering.
    (d) reflection and absorption.

The answer is (c)

Problem–71

The total cross section of a target atom is made up of

    (a) the absorption cross section.
    (b) the scattering cross section.
    (c) the absorption and scattering cross sections.
    (d) the reflection and absorption cross sections.

The answer is (c)

Problem–72

All of the following are phenomena of neutron interactions except

    (a) inelastic scattering.
    (b) elastic scattering.
    (c) fission.
    (d) fusion.

The answer is (d)

Problem–73

All of the following are words used to describe neutron kinetic energy levels except

    (a) cold.
    (b) thermal.
    (c) slow.
    (d) freezing.

The answer is (d)

Problem–74

All of the following are words used to describe neutron kinetic energy levels except

    (a) slow (resonant).
    (b) fast.
    (c) supersonic.
    (d) relativistic.

The answer is (c)

Problem–75

The reduction of nuclear radiation intensity (called attenuation) is accomplished by

    (a) matter.
    (b) antimatter.
    (c) shielding.
    (d) neurons.

The answer is (c)

Problem–76

The thickness of material required to attenuate radiation to a particular level depends on

    (a) the particle.
    (b) the particle energy.
    (c) the shielding material.
    (d) the particle, its energy, and the shielding material.

The answer is (d)

Problem–77

Particles that are easily stopped within a few millimeters because their double charges generate path ionization and because they are susceptible to electrostatic interaction are

  (a) slow neutrons.
  (b) alpha radiation.
  (c) beta radiation.
  (d) gamma radiaton.

The answer is (b)

Problem–78

Radiation consisting of singly charged particles that penetrate to intermediate distances are called

  (a) fast neutrons.
  (b) alpha radiation.
  (c) beta radiation.
  (d) gamma radiaton.

The answer is (c)

Problem–79

Radiation with no charge, which produces no ionization, and which is difficult to attenuate thus posing a major health threat is

  (a) slow neutrons.
  (b) alpha radiation.
  (c) beta radiation.
  (d) gamma radiation.

The answer is (d)

Problem–80

The ability of a substance to absorb neutrons depends upon its

  (a) absorption cross section.
  (b) scattering cross section.
  (c) total cross section.
  (d) atomic number.

The answer is (a)

Problem–81

All of the following materials can absorb neutrons well because of their high-absorption cross section, σ_a, except

  (a) cadmium.
  (b) boron.
  (c) graphite.
  (d) helium.

The answer is (d)

Problem–82

Gamma attentuation is affected by

  (a) the photoelectric effect.
  (b) pair production.
  (c) Compton scattering.
  (d) the photoelectric effect, pair production, and Compton scattering.

The answer is (d)

Problem–83

The amount of a radiation shield's dimensional geometry that reduces radiation to half its original value is called the

  (a) half-value mass.
  (b) half-value thickness.
  (c) semicross section.
  (d) logarithmic decrement.

The answer is (b)

Problem–84

The amount of a radiation shield's density that reduces radiation to half its original value is called the

  (a) half-value mass.
  (b) half-value thickness.
  (c) semicross section.
  (d) logarithmic decrement.

The answer is (a)

Problem–85

Radiation exposure, the measure of gamma radiation at the surface of an object, is measured in

(a) rems.
(b) rads.
(c) roentgens.
(d) roentgens per second.

The answer is (c)

Problem–86

Radiation exposure rate, the rate of gamma radiation at the surface of an object, is measured in

(a) rems.
(b) rads.
(c) roentgens.
(d) roentgens per second.

The answer is (d)

Problem–87

Exposure is a measure of ionization surrounding a person, but biological damage is dependent on the amount of energy

(a) striking the surface.
(b) passing through the body.
(c) absorbed.
(d) reflected by the surface.

The answer is (c)

Problem–88

When a nucleus splits into smaller fragments the process is called

(a) fusion.
(b) fission.
(c) the photoelectric effect.
(d) the Compton effect.

The answer is (b)

Problem–89

When two or more light atoms have sufficient energy (available only at high temperatures and velocities) to fuse together to form a heavier nucleus the process is called

(a) fusion.
(b) fission.
(c) the photoelectric effect.
(d) the Compton effect.

The answer is (a)

Problem–90

During the fusion process, mass is lost and converted to energy according to

(a) the Heisenburg uncertainty principle.
(b) the Compton effect.
(c) Einstein's law.
(d) the second law of thermodynamics.

The answer is (c)

Problem–91

The most highly developed device for confining plasmas with magnetic fields is the

(a) Tokamak.
(b) Tomahawk.
(c) breeder reactor.
(d) cyclotron.

The answer is (a)

Problem–92

All of the following are practical applications of Einstein's principle of special relativity except

(a) mass increase.
(b) length contraction.
(c) time dilation.
(d) space warping.

The answer is (d)

Problem–93

The postulate that no signal or energy can be transmitted with a speed greater than the speed of light is consistent with

    (a) the Heisenburg uncertainty principle.
    (b) the Compton effect.
    (c) Einstein's law.
    (d) Newton's second law.

The answer is (c)

Problem–94

The total energy of an electron in the same shell is defined by the

    (a) principal quantum number.
    (b) azimuthal quantum number.
    (c) magnetic quantum number.
    (d) Hund rule.

The answer is (a)

Problem–95

The direction of an electron's angular momentum vector is defined by the

    (a) principal quantum number.
    (b) azimuthal quantum number.
    (c) magnetic quantum number.
    (d) electron spin quantum number.

The answer is (c)

Problem–96

The electron's spin angular momentum vector is defined by the

    (a) azimuthal quantum number.
    (b) magnetic quantum number.
    (c) electron spin quantum number.
    (d) Hund rule.

The answer is (c)

Problem–97

The magnitude of an electron's angular momentum vector is defined by the

    (a) principal quantum number.
    (b) azimuthal quantum number.
    (c) electron spin quantum number.
    (d) Hund rule.

The answer is (b)

Problem–98

The fact that each orbital of a set of equal-energy orbitals must be occupied with an electron before any orbital has two electrons is specified by the

    (a) principal quantum number.
    (b) azimuthal quantum number.
    (c) magnetic quantum number.
    (d) Hund rule.

The answer is (d)

Problem–99

The statement that no two electrons can have the same set of four quantum numbers is known as the

    (a) Hund rule.
    (b) Heisenburg uncertainty principle.
    (c) Pauli exclusion principle.
    (d) Schrödinger equation.

The answer is (c)

Problem–100

All of the following terms describe the radiation lines from transitions of electrons in an atom except

    (a) sharp.
    (b) principal.
    (c) obtuse.
    (d) fundamental.

The answer is (c)

Chapter 8
Materials Science/Structure of Matter

Refer to the following illustration for Probs. 1 through 9.

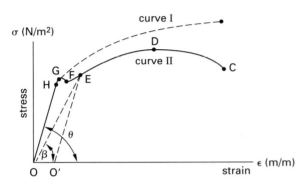

Problem–1

For the stress-strain curve of carbon steel shown, curve I is a plot of

(a) a true stress-strain curve.
(b) an engineering stress-strain curve.
(c) tensile force versus elongation.
(d) the modulus of elasticity.

The answer is (a)

Problem–2

The slope of the line OH is

(a) a true stress-strain curve.
(b) an engineering stress-strain curve.
(c) tensile force versus elongation.
(d) the modulus of elasticity.

The answer is (d)

Problem–3

For the stress-strain curve of carbon steel shown, curve II is a plot of

(a) a true stress-strain curve.
(b) an engineering stress-strain curve.
(c) tensile force versus elongation.
(d) the modulus of elasticity.

The answer is (b)

Problem–4

The slope of the line OE is the

(a) secant modulus.
(b) proportionality limit.
(c) yield point.
(d) ultimate strength.

The answer is (a)

Problem–5

Point H indicates the

(a) secant modulus.
(b) proportionality limit.
(c) yield point.
(d) ultimate strength.

The answer is (b)

Problem–6

Point F indicates the

(a) proportionality limit.
(b) yield point.
(c) ultimate strength.
(d) the permanent set.

The answer is (b)

Problem–7

Point O′ indicates the

    (a) proportionality limit.
    (b) yield point.
    (c) ultimate strength.
    (d) permanent set.

The answer is (d)

Problem–8

Point D indicates the

    (a) fracture point.
    (b) yield point.
    (c) ultimate strength.
    (d) permanent set.

The answer is (c)

Problem–9

Point C indicates the

    (a) fracture point.
    (b) elastic limit.
    (c) point.
    (d) the ultimate strength.

The answer is (a)

Problem–10

The area X represents

    (a) the modulus of elasticity.
    (b) the modulus of resilience.
    (c) the modulus of toughness.
    (d) Young's modulus.

The answer is (b)

Problem–11

The area Y represents

    (a) the modulus of elasticity.
    (b) the modulus of resilience.
    (c) the modulus of toughness.
    (d) Young's modulus.

The answer is (c)

Problem–12

Poisson's ratio is the ratio of

    (a) true stress to true strain.
    (b) lateral strain to axial strain.
    (c) shear stress to shear strain.
    (d) tensile force to elongation.

The answer is (b)

Refer to the following illustration for Probs. 10 and 11.

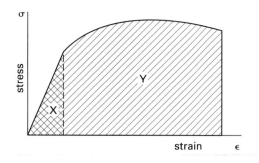

Problem-13

For the three types of tensile fracture shown, rank them in order of increasing ductility, that is, from the most brittle to the most ductile material.

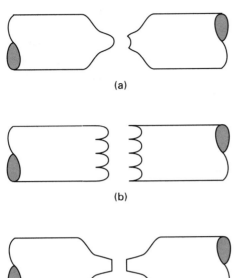

(a)

(b)

(c)

(a) a, b, c
(b) b, c, a
(c) b, a, c
(d) c, a, b

The answer is (c)

Problem-14

Most engineering work is based on engineering stress and strain (rather than true stress and strain) because

(a) engineering stress and strain is more accurate.
(b) engineering stress and strain is easier to use.
(c) design using ductile materials is limited to the elastic region where there is little difference.
(d) reduction of cross-sectional area of parts at their service stresses is well known due to Poisson's ratio, but it is a needless complication.

The answer is (c)

Problem-15

Percent elongation is a measure of

(a) the ratio of ultimate failure strain to yielding strain.
(b) energy per unit volume stored in a deformed material.
(c) total plastic strain at failure.
(d) strain energy per unit volume to reach the yield point.

The answer is (c)

Problem-16

Resilience (elastic toughness) is a measure of

(a) energy per unit volume stored in a deformed material.
(b) total plastic strain at failure.
(c) strain energy per unit volume to reach the yield point.
(d) strain energy (work per unit volume) to cause fracture.

The answer is (c)

Problem-17

Toughness is a measure of

(a) the ratio of ultimate failure strain to yielding strain.
(b) energy per unit volume stored in a deformed material.
(c) strain energy per unit volume to reach the yield point.
(d) strain energy (work per unit volume) to cause fracture.

The answer is (d)

Problem–18

Strain energy is a measure of

(a) the ratio of ultimate failure strain to yielding strain.
(b) energy per unit volume stored in a deformed material.
(c) total plastic strain at failure.
(d) strain energy per unit volume to reach the yield point.

The answer is (b)

Problem–19

Ductility is a measure of

(a) the ratio of ultimate failure strain to yielding strain.
(b) energy per unit volume stored in a deformed material.
(c) total plastic strain at failure.
(d) strain energy per unit volume to reach the yield point.

The answer is (a)

Problem–20

Fatigue testing determines

(a) failure of ductile materials in tension or compression below the elastic limit.
(b) failure in torsion.
(c) capacity of a surface to resist deformation.
(d) failure after repeated loadings even if the stress level never exceeds the material's ultimate strength.

The answer is (d)

Problem–21

Hardness testing determines the

(a) failure of brittle materials in compression.
(b) failure of ductile materials in tension or compression below the elastic limit.
(c) capacity of a surface to resist deformation.
(d) failure after repeated loadings even if the stress level never exceeds the material's ultimate strength.

The answer is (c)

Problem–22

Compression testing determines the

(a) failure of brittle materials in compression.
(b) failure of ductile materials in tension or compression below the elastic limit.
(c) capacity of a surface to resist deformation.
(d) failure after repeated loadings even if the stress level never exceeds the material's ultimate strength.

The answer is (a)

Problem–23

Tensile testing determines

(a) failure of ductile materials in tension or compression below the elastic limit.
(b) failure in torsion.
(c) capacity of a surface to resist deformation.
(d) failure after repeated loadings even if the stress level never exceeds the material's ultimate strength.

The answer is (a)

Problem–24

Torsion testing determines

    (a) failure of brittle materials in compression.

    (b) failure of ductile materials in tension or compression below the elastic limit.

    (c) failure in torsion.

    (d) capacity of a surface to resist deformation.

The answer is (c)

Problem–25

All of the following are hardness tests except

    (a) Brinnel tests.

    (b) Rockwell tests.

    (c) Meyer-Vickers tests.

    (d) Charpy tests.

The answer is (d)

Problem–26

What is the hardest form of steel?

    (a) bainite

    (b) ferrite

    (c) pearlite

    (d) martensite

The answer is (d)

Problem–27

All of the following activities increase the fatigue endurance limit of a metal except

    (a) polishing.

    (b) cold working.

    (c) surface hardening.

    (d) filleting joints.

The answer is (b)

Problem–28

Which of the following processes can increase the deformation resistance of steel?

    (I) hot working

    (II) adding alloying elements

    (III) tempering

    (IV) hardening

    (a) I and II

    (b) I and III

    (c) II and IV

    (d) III and IV

The answer is (c)

Problem–29

Consider the Mohs Hardness Scale shown (Mohs number as a function of Vickers or Knoop indentation hardness, kg/mm^2), and the following materials.

    (I) topaz

    (II) fluorite (fluorspar)

    (III) diamond

    (IV) orthoclase (feldspar)

    (V) talc

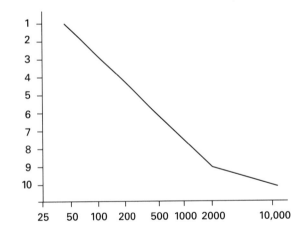

Arrange the materials in order of increasing hardness.

    (a) I, II, III, V, IV

    (b) IV, V, II, I, III

    (c) V, II, IV, I, III

    (d) III, II, V, IV, I

The answer is (c)

Problem–30

What test is used to determine the toughness of a material?

(a) impact
(b) hardness
(c) fatigue
(d) creep

The answer is (a)

Problem–31

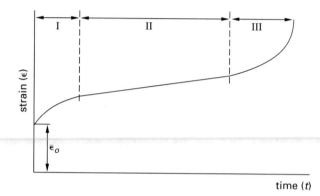

In the plot for creep data shown, the distinct differences in slope between regions I, II, and III are due primarily to the interplay between which of the following?

(a) cold working and fatigue
(b) temperature and stress
(c) stress and time
(d) strain hardening and annealing

The answer is (d)

Problem–32

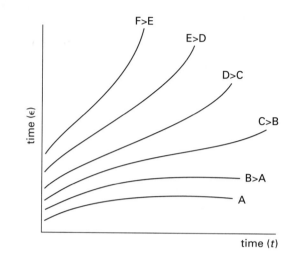

The increased creep rate between curves in the creep rate plot shown is due to increasing which of the following?

(a) number of cycles
(b) temperature
(c) stress
(d) strain

The answer is (c)

Problem–33

A strong material

(a) shows high ultimate strength.
(b) yields greatly before breaking.
(c) yields little before breaking.
(d) exhibits a high modulus of elasticity.

The answer is (a)

Problem–34

A tough material

(a) shows low ultimate strength.
(b) shows high ultimate strength.
(c) yields greatly before breaking.
(d) exhibits a high modulus of elasticity.

The answer is (c)

Problem–35

A brittle material

    (a) shows low ultimate strength.
    (b) yields greatly before breaking.
    (c) yields little before breaking.
    (d) exhibits a high modulus of elasticity.

The answer is (c)

Problem–36

A hard material

    (a) shows low ultimate strength.
    (b) shows high ultimate strength.
    (c) yields little before breaking.
    (d) exhibits a high modulus of elasticity.

The answer is (d)

Problem–37

A weak material

    (a) shows low ultimate strength.
    (b) shows high ultimate strength.
    (c) exhibits a low modulus of elasticity.
    (d) yields little before breaking.

The answer is (a)

Problem–38

A soft material

    (a) shows low ultimate strength.
    (b) shows high ultimate strength.
    (c) exhibits a low modulus of elasticity.
    (d) yields little before breaking.

The answer is (c)

Refer to the following illustrations for Probs. 39 through 43.

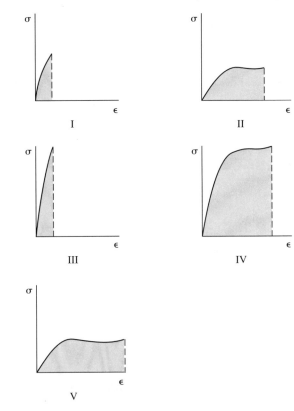

Problem–39

In the stress-strain diagrams shown, material I is

    (a) strong and tough.
    (b) hard and strong.
    (c) soft and weak.
    (d) weak and brittle.

The answer is (d)

Problem–40

In the stress-strain diagrams shown, material II is

    (a) strong and tough.
    (b) hard and strong.
    (c) soft and weak.
    (d) weak and brittle.

The answer is (c)

Problem–41

In the stress-strain diagrams shown, material III is

    (a) strong and tough.
    (b) hard and strong.
    (c) soft and weak.
    (d) weak and brittle.

The answer is (b)

Problem–42

In the stress-strain diagrams shown, material IV is

    (a) strong and tough.
    (b) hard and strong.
    (c) soft and weak.
    (d) weak and brittle.

The answer is (a)

Problem–43

In the stress-strain diagrams shown, material V is

    (a) strong and tough.
    (b) soft and weak.
    (c) weak and brittle.
    (d) weak and tough.

The answer is (d)

Problem–44

All of the following increase strength and reduce ductility of metals except

    (a) alloying materials.
    (b) imperfections.
    (c) free movement of dislocations.
    (d) strain hardening.

The answer is (c)

Problem–45

At a specific temperature, the phase of a material will have all of the following distinct characteristics except

    (a) composition.
    (b) crystalline structure.
    (c) magnetic properties.
    (d) freezing point.

The answer is (d)

Problem–46

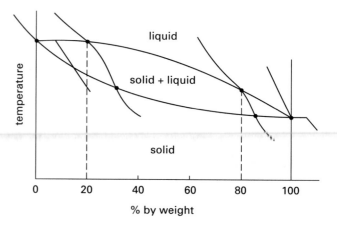

In the sketch of the temperature-time curves of a solidifying alloy, the locus (loci) of inflection points of the curve of transition temperatures is (are) which of the following?

    (a) the solidus line
    (b) the liquidus line
    (c) the eutectic points
    (d) the solidus and liquidus lines

The answer is (d)

Problem–47

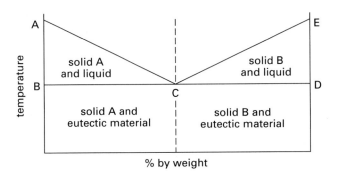

% by weight

The point at which the two components of the binary alloy are perfectly miscible is which of the following?

(a) point A
(b) point B
(c) point C
(d) point D

The answer is (c)

Refer to the following illustration for Probs. 48 through 50.

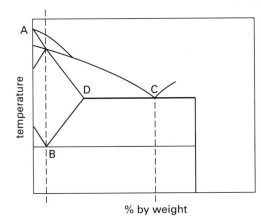

% by weight

Problem–48

In the accompanying phase diagram, the eutectic reaction is shown at point

(a) A.
(b) B.
(c) C.
(d) D.

The answer is (c)

Problem–49

In the accompanying phase diagram, a peritectic reaction is shown at point

(a) A.
(b) B.
(c) C.
(d) D.

The answer is (a)

Problem–50

In the accompanying phase diagram, the eutectoid reaction is shown at point

(a) A.
(b) B.
(c) C.
(d) D.

The answer is (b)

Problem–51

All of the following are differing characteristics of allotropes except

(a) atomic microstructure.
(b) volume.
(c) composition.
(d) electrical resistance.

The answer is (c)

Problem–52

Materials with the same composition but different atomic structures are

    (a) isotopes.
    (b) isomers.
    (c) allotropes.
    (d) polymers.

The answer is (c)

Problem–53

All of the following reaction points describe the transition of phases in a binary alloy except a(n)

    (a) eutectic reaction.
    (b) eutectoid reaction.
    (c) austentite reaction.
    (d) peritectic reaction.

The answer is (c)

Problem–54

All of the following methods are used to surface harden steel except

    (a) carburizing (cementation).
    (b) cyaniding.
    (c) flame hardening.
    (d) full annealing.

The answer is (d)

Problem–55

Surface hardening is desirable for a product that is subject to

    (a) high tensile loads.
    (b) high shear loads.
    (c) fatigue.
    (d) high impact loads.

The answer is (c)

Problem–56

In the manufacture of steel, all of the following are processes to relieve internal stresses, refine grain size, and soften material (i.e., improve machinability) except

    (a) full annealing.
    (b) tempering.
    (c) normalizing.
    (d) quenching.

The answer is (d)

Problem–57

The division between cast iron and steel in iron-carbon alloys is approximately what percentage of carbon?

    (a) 0.1%
    (b) 0.7%
    (c) 2.0%
    (d) 4.4%

The answer is (c)

Problem–58

The temperature below which a steel alloy becomes magnetic is called the

    (a) triple point.
    (b) Curie point.
    (c) freezing point.
    (d) the lower critical point.

The answer is (b)

Problem–59

The purpose of quenching steel is to

    (a) relieve stress.
    (b) control hardness and ductility of a heated material.
    (c) increase grain size.
    (d) increase strength and lower ductility.

The answer is (b)

Problem–60

The aggregation of atoms in a compound is more stable than the individual atoms because

(a) nature abhors a vacuum.
(b) the energy of the aggregation is lower than that of the individual atoms.
(c) the energy of the aggregation is higher than that of the individual atoms.
(d) atoms rarely occur alone in nature.

The answer is (b)

Problem–61

A crystal can maintain its equilibrium position because

(a) all angles in the crystal lattice are 90°.
(b) the repulsion forces of the inner electrons balance the attraction of the positive and negative charges of the ions.
(c) all sides of the lattice are of equal length giving a cubic crystal.
(d) all sides of the lattice are of different length giving a hexagonal lattice.

The answer is (b)

Problem–62

How many different crystal structures are found in nature?

(a) 12
(b) 14
(c) 16
(d) 20

The answer is (b)

Problem–63

All of the following characteristics identify the more common crystalline lattice structures except the

(a) length of the lattice sides.
(b) locations of atoms within lattice faces.
(c) equivalent spherical angles between lattice corners and faces.
(d) locations of atoms inside the lattice body.

The answer is (c)

Problem–64

In a crystalline structure, the packing factor is the

(a) distance between lattice atoms.
(b) number of closest touching atoms.
(c) ratio of the volume of the atoms to the volume of the cell.
(d) ratio of cell mass to cell volume.

The answer is (c)

Problem–65

In a crystalline structure, the number of atoms in the cell is the

(a) distance between lattice atoms.
(b) number of closest touching atoms.
(c) ratio of the volume of the atoms to the volume of the cell.
(d) number of lattice points.

The answer is (d)

Problem–66

In a crystalline structure, the lattice constant is the

(a) distance between lattice atoms.
(b) number of closest touching atoms.
(c) number of lattice points.
(d) ratio of cell mass to cell volume.

The answer is (a)

Problem-67

In a crystalline structure, the density of the ionic solid is the

(a) distance between lattice atoms.
(b) number of closest touching atoms.
(c) number of lattice points.
(d) ratio of cell mass to cell volume.

The answer is (d)

Problem-68

In a crystalline structure, the coordination number is the

(a) distance between lattice atoms.
(b) number of closest touching atoms.
(c) number of lattice points.
(d) ratio of cell mass to cell volume.

The answer is (b)

Problem-69

What is the strongest type of bond?

(a) ionic
(b) covalent
(c) van der Waals
(d) metallic

The answer is (b)

Problem-70

Most metallic crystals form one of the following three lattice structures.

(a) hexagonal close-packed, simple tetragonal, or cubic
(b) base-centered orthorhombic, body-centered orthorhombic, or rhombohedral
(c) body-centered cubic, face-centered cubic, or hexagonal close-packed
(d) simple tetragonal, body-centered tetragonal, or body-centered cubic

The answer is (c)

Problem-71

The planes of a crystalline lattice can be specified by

(a) Burger's vectors.
(b) Taylor-Orowan dislocations.
(c) Fick's law.
(d) Miller indices.

The answer is (d)

Problem-72

All of the following are crystalline point defects except

(a) Schottky defects.
(b) interstitial impurity atoms.
(c) screw dislocations.
(d) vacancies.

The answer is (c)

Problem-73

All of the following are structure-sensitive properties affected by crystalline imperfections and defects except

(a) density.
(b) electrical conductivity.
(c) yield strength.
(d) ultimate strength.

The answer is (a)

Problem-74

All of the following are structure-insensitive quantities that are not affected by crystalline imperfections and defects except

(a) semiconductor properties.
(b) melting point.
(c) specific heat.
(d) shear modulus.

The answer is (a)

Problem–75

The shell structure and dimensions of crystals can be determined by

  (a) precision metrology.
  (b) x-ray diffraction.
  (c) interferometry.
  (d) Schlieren photographs.

The answer is (b)

Problem–76

In studying the x-ray diffraction of crystal planes, the order of diffraction or order of deflection, n, is given by the expression

$$n\lambda = 2d_{hkl}\sin\theta$$

λ is the x-ray wavelength, d_{hkl} is the interplanar spacing, and θ is the incident beam angle. This equation is known as

  (a) Bragg's law.
  (b) Fick's law.
  (c) Hooke's law.
  (d) Boyles' law.

The answer is (a)

Problem–77

The most common type of crystal defects are

  (a) point defects.
  (b) line defects.
  (c) grain boundary defects.
  (d) surface defects.

The answer is (b)

Problem–78

The quantity needed to close the circuit around a crystalline dislocation is called

  (a) a gradient vector.
  (b) a unit vector.
  (c) Burgers' vector.
  (d) a Poynting vector.

The answer is (c)

Problem–79

The Burgers' vector is similar to

  (a) the unit normal to a crystalline surface of complex curvature.
  (b) the error of closure that surveyors encounter when they survey around an area.
  (c) the direction of steepest descent on a crystalline surface.
  (d) a vector cross product between crystalline surface normals.

The answer is (b)

Problem–80

All of the following parameters affect the movement of point defects within a crystal except

  (a) the material.
  (b) temperature.
  (c) atmospheric pressure.
  (d) defect concentration.

The answer is (c)

Problem–81

Crystalline structures are stable because they

  (a) allow diffusion of point defects by Fick's law.
  (b) permit atoms to exist at lower energy states than would be experienced in the absence of crystal formation.
  (c) give rise to the existence of Burgers' vector.
  (d) allow dislocation under application of heat or shear stress.

The answer is (b)

Problem–82

The imperfections in a crystal give rise to

    (a) disorder in the lattice and a higher energy state. (The higher energy state is known as strain energy.)
    (b) disorder in the lattice and a lower energy state.
    (c) opposite sides of the crystalline lattice of unequal length.
    (d) diamonds of outstanding clarity and brilliance.

The answer is (a)

Problem–83

The combination of the preferred slip plane and the slip direction is known as a

    (a) crystalline lattice.
    (b) slip system.
    (c) point defect.
    (d) lattice direction.

The answer is (b)

Problem–84

All of the following are typical crystalline defects except

    (a) point defects.
    (b) line defects.
    (c) grain boundary defects.
    (d) mass defects.

The answer is (d)

Problem–85

The spontaneous and permanent alignment of electrons creating a strongly magnetic material in such materials as alpha-iron, cobalt, and nickel is known as

    (a) ferrimagnetism.
    (b) ferromagnetism.
    (c) paramagnetism.
    (d) diamagnetism.

The answer is (b)

Problem–86

The weakly repulsive force experienced in most non-metals and organic compounds (e.g., bismuth, paraffin, and silver) when exposed to a magnetic field is

    (a) ferrimagnetism.
    (b) paramagnetism.
    (c) diamagnetism.
    (d) antiferromagnetism.

The answer is (c)

Problem–87

The weak attraction experienced by most alkali and transition metals when exposed to an external magnetic field is

    (a) ferrimagnetism.
    (b) paramagnetism.
    (c) diamagnetism.
    (d) antiferromagnetism.

The answer is (b)

Problem–88

The strong magnetism that occurs in certain ceramic (crystalline) compounds such as ferrite (e.g., $MnFe_2O_4$ and $ZnFe_2O_4$), spinels ($MgAl_2O_4$) and garnets is known as

    (a) ferrimagnetism.
    (b) ferromagnetism.
    (c) diamagnetism.
    (d) antiferromagnetism.

The answer is (a)

Problem–89

The weak attraction to a magnet that is typical of salts of the transition elements and elements of aluminum, copper, gold, and lead is known as

    (a) ferrimagnetism.
    (b) paramagnetism.
    (c) diamagnetism.
    (d) antiferromagnetism.

The answer is (d)

Problem-90

Materials whose magnetic domains remain aligned even after the external magnetic field is removed are called

(a) magnetically hard.
(b) magnetically soft.
(c) magnetically neutral.
(d) antiferromagnetic.

The answer is (a)

Problem-91

An index of the energy lost in a complete cycle of magnetization is given by

(a) permeability.
(b) the hysteresis loop.
(c) susceptibility.
(d) paramagnetism.

The answer is (b)

Refer to the following illustrations for Probs. 92 and 93.

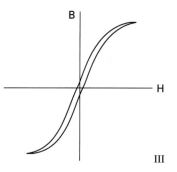

III

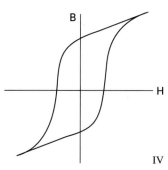

IV

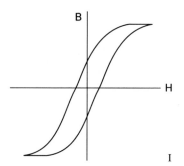

I

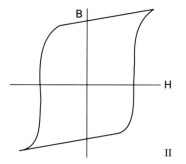

II

Problem-92

Arrange the materials whose magnetic characteristics are shown in order of hardness, from soft to hard.

(a) I, II, IV, III
(b) II, IV, I, III
(c) IV, III, I, II
(d) III, I, IV, II

The answer is (d)

Problem-93

For the magnets whose hysteresis loops are shown in the sketch, which one has the most lifting power?

(a) I
(b) II
(c) III
(d) IV

The answer is (b)

Problem–94

Steel is the most prevalent engineering metal for all of the following reasons except

(a) the abundance of iron ore.
(b) its low density.
(c) simplicity of production.
(d) predictability of performance.

The answer is (b)

Problem–95

All of the following properties are characteristic of metals except

(a) high thermal conductivity.
(b) high chemical reactivity.
(c) high ductility.
(d) electromagnetic transparency.

The answer is (d)

Problem–96

The subject that covers the refinement of pure metals from their ores is

(a) ceramic toughening.
(b) alchemy.
(c) extractive metallurgy.
(d) composite refining.

The answer is (c)

Problem–97

All of the following ingredients are used as steel alloying ingredients to increase strength, hardness, or toughness except

(a) carbon.
(b) chromium.
(c) copper.
(d) nickel.

The answer is (c)

Problem–98

Aluminum is particularly satisfactory for applications requiring all of the following characteristics except

(a) low weight.
(b) high strength.
(c) good electrical conductivity.
(d) good thermal conductivity.

The answer is (b)

Problem–99

The material characteristic of low thermal expansion is particularly desirable in

(a) screw machine parts.
(b) surveyors' tapes.
(c) gears.
(d) cutting tools.

The answer is (b)

Problem–100

The three main performance requirements for tool steel are

(a) strength, ductility, and thermal conductivity.
(b) toughness, wear resistance, and hot hardness.
(c) radiation resistance, ductility, and high thermal conductivity.
(d) chemical resistance, toughness, and strength.

The answer is (b)

Problem–101

All of the following are alloys of nickel except

(a) K-monel.
(b) inconel-X.
(c) chromium molybdenum steel.
(d) nichrome.

The answer is (c)

Problem–102

Vulcanization of rubber is accomplished by heating raw rubber with small amounts of

   (a) carbon black.
   (b) sulfur.
   (c) silver.
   (d) paraffin.

The answer is (b)

Problem–103

The ratio of the strength of dry wood to that of wet or green wood is approximately

   (a) 2.
   (b) 4.
   (c) 8.
   (d) 20.

The answer is (a)

Problem–104

The ratio of longitudinal strength to cross-grain strength for wood may be as much as

   (a) 4.
   (b) 8.
   (c) 20.
   (d) 40.

The answer is (d)

Problem–105

All of the following are examples of ceramics except

   (a) brick.
   (b) Portland cement.
   (c) paraffin.
   (d) refractories.

The answer is (c)

Problem–106

Ceramic materials are marked by all of the following characteristics except

   (a) the absence of free valence bonds.
   (b) crystalline structure.
   (c) high compressive strength.
   (d) high tensile strength.

The answer is (d)

Problem–107

The absence of free electrons in ceramics make them

   (a) good electrical conductors.
   (b) poor electrical conductors.
   (c) good thermal conductors.
   (d) materials with a high coefficient of thermal expansion.

The answer is (b)

Problem–108

Compounds with the same chemical formulas but different physical structures are called

   (a) isomers.
   (b) isotopes.
   (c) polygons.
   (d) polymorphs.

The answer is (d)

Problem–109

Polymorphism is demonstrated by

   (a) different molecular weights.
   (b) different arrangements of the same atoms.
   (c) crystals of the same materials with differing numbers of sides.
   (d) different crystalline structures at different temperatures.

The answer is (d)

Problem–110

All of the following are common constituents of concrete except

  (a) Portland cement.
  (b) sand.
  (c) aggregates.
  (d) calcium chloride.

The answer is (d)

Problem–111

The ratio of 1:2:3 for a concrete mixture gives the relative weights of which of the following constituents?

  (a) cement, coarse aggregate, and water
  (b) cement, sand, and water
  (c) cement, fine aggregate, and coarse aggregate
  (d) water, fine aggregate, and coarse aggregate

The answer is (c)

Problem–112

The strength of concrete is determined by its

  (a) ultimate tensile strength.
  (b) ultimate compressive strength.
  (c) ultimate shear strength.
  (d) secant modulus of elasticity.

The answer is (b)

Problem–113

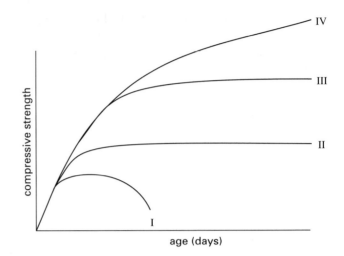

The compressive strength of a moistured-cured concrete is given by which of the following curves?

  (a) I
  (b) II
  (c) III
  (d) IV

The answer is (d)

Problem–114

The addition of steel to concrete is primarily to

  (a) increase material density.
  (b) increase torsional strength.
  (c) increase tensile strength.
  (d) increase abrasion resistance.

The answer is (c)

Problem–115

Steel used as a wire mesh in concrete is for

  (a) increased material density.
  (b) increased yield strength.
  (c) shrinkage and expansion control.
  (d) increased abrasion resistance.

The answer is (c)

Problem—116

The use of admixtures (hydrated lime, fly ash, etc.) with concrete is to

(a) reduce weight.
(b) increase durability.
(c) improve workability, hardening, or strength characteristics.
(d) accelerate curing.

The answer is (c)

Problem—117

Surface hardening to produce a hard outer surface with a ductile interior is desirable for metal (steel) products subjected to

(a) high tensile loads.
(b) shock.
(c) fatigue.
(d) high compressive loads.

The answer is (c)

Problem—118

All of the following are processes used to surface-harden steel except

(a) carburizing.
(b) cyaniding.
(c) flame hardening.
(d) tempering.

The answer is (d)

Problem—119

Which of the following is the only allotropic alloy of steel?

(a) aluminum alloys
(b) copper alloys
(c) magnesium
(d) iron-carbon alloys

The answer is (d)

Problem—120

During the quenching process in heat treating steel, agitation of the fluids used in the quenching has what effect on the severity of the quenching?

(a) increases quenching severity
(b) decreases quenching severity
(c) has no effect on quenching severity
(d) reverses quenching severity

The answer is (a)

Problem—121

All of the following materials are used in the quenching process for steel manufacturing except

(a) air.
(b) brine.
(c) hydrogen.
(d) water.

The answer is (c)

Problem—122

All of the following are characteristics of coarse-grained structures except

(a) less toughness.
(b) less ductility.
(c) more machinability.
(d) less corrosion resistance.

The answer is (d)

Problem—123

The purpose of recrystallization of all metals is to

(a) increase hardness and ductility.
(b) increase the amount of alloying ingredient absorbed by the solid mixture.
(c) relieve stresses induced during cold working.
(d) reduce the temperature of the mixture below the eutectic line.

The answer is (c)

Problem–124

Processes to relieve the internal stresses in the manufacture of steel include all of the following except

    (a) full annealing processes.
    (b) normalizing processes.
    (c) tempering processes.
    (d) carburizing processes.

The answer is (d)

Problem–125

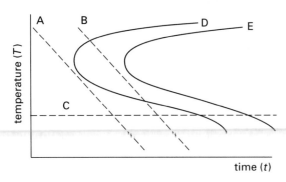

A tempering process is shown by which of the following curves?

    (a) A
    (b) B
    (c) C
    (d) D

The answer is (c)

Problem–126

Allotropes have the same chemical composition but different

    (a) crystalline structures.
    (b) densities.
    (c) electrical resistance.
    (d) all of the above

The answer is (d)

Problem–127

Hardness tests measure

    (a) ultimate strength.
    (b) toughness.
    (c) capacity of a surface to resist deformation.
    (d) fatigue strength.

The answer is (c)

Problem–128

All of the following are destructive hardness tests except

    (a) Mohs tests.
    (b) file hardness tests.
    (c) Shore hardness tests.
    (d) cutting hardness tests.

The answer is (c)

Problem–129

All of the following are nondestructive tests except

    (a) ultrasonic tests.
    (b) infrared inspection tests.
    (c) Shore hardness tests.
    (d) Mohs tests.

The answer is (d)

Problem–130

Which of the following nondestructive methods can be used to find defects in a manufactured steel part?

    (a) ultrasound
    (b) die-penetrant
    (c) magnetic particle
    (d) all of the above

The answer is (d)

Chapter 9

Electric Circuits

Problem–1

Static electricity can be generated by rubbing the following pairs of materials together except

(a) silk on a glass rod.
(b) fur on rubber.
(c) one wooden stick on another.
(d) fur on a plastic rod.

The answer is (c)

Problem–2

The charge on an electron is

(a) 1 electrostatic unit.
(b) 1 faraday.
(c) 1 ohm.
(d) 1 ampere.

The answer is (a)

Problem–3

In a parallel plate capacitor, the charge Q is proportional to

(a) the reciprocal of capacitance squared.
(b) the reciprocal of capacitance.
(c) capacitance.
(d) capacitance squared.

The answer is (c)

Problem–4

What is the total capacitance of a collection of capacitors in series?

(a) $\dfrac{1}{C} = \dfrac{1}{C_1} + \dfrac{1}{C_2} + \ldots \dfrac{1}{C_n}$
(b) $C = C_1 + C_2 + \ldots C_n$
(c) $C = C_1 C_2 + C_2 C_3 + \ldots (C_n - 1)(C_n)$
(d) $C = \dfrac{1}{C_1 C_2} + \dfrac{1}{C_2 C_3} + \ldots \dfrac{1}{(C_n - 1)(C_n)}$

The answer is (a)

Problem–5

What is the total capacitance of a collection of capacitors in parallel?

(a) $\dfrac{1}{C} = \dfrac{1}{C_1} + \dfrac{1}{C_2} + \ldots \dfrac{1}{C_n}$
(b) $C = C_1 + C_2 + \ldots C_n$
(c) $C = C_1 C_2 + C_2 C_3 + \ldots (C_n - 1)(C_n)$
(d) $C = \dfrac{1}{C_1 C_2} + \dfrac{1}{C_2 C_3} + \ldots \dfrac{1}{(C_n - 1)(C_n)}$

The answer is (b)

Problem–6

The property of a device that impedes current flow is

(a) capacitance.
(b) inductance.
(c) resistance.
(d) permittivity.

The answer is (c)

Problem–7

The property of a device in which a magnetic field impedes changes in current flow is

(a) capacitance.
(b) inductance.
(c) resistivity.
(d) permittivity.

The answer is (b)

Problem–8

The property of a material that defines capacitance is

(a) conductance.
(b) inductance.
(c) resistivity.
(d) permittivity.

The answer is (d)

Problem–9

The property of a device that quantifies how well it stores electric charge is

(a) capacitance.
(b) inductance.
(c) resistivity.
(d) permittivity.

The answer is (a)

Problem–10

The property of a material, which along with the geometry of the device characterizes its resistance is

(a) capacitance.
(b) inductance.
(c) resistivity.
(d) permittivity.

The answer is (c)

Problem–11

Electrical resistance of a device is a function of

(a) device configuration.
(b) material property and configuration.
(c) current.
(d) voltage.

The answer is (b)

Problem–12

The reciprocal of the resistance R is the

(a) impedance L.
(b) capacitance C.
(c) conductance G.
(d) resistivity ρ.

The answer is (c)

Problem–13

A circular mil (abbreviated "cmil") represents the

(a) impedance L.
(b) magnetic flux.
(c) cross-sectional area of a circular conductor.
(d) voltage drop of a linear resistor.

The answer is (c)

Problem–14

A resistor is characterized by a resistivity ρ, an area A, and a length l. The resistance of the device is

(a) directly proportional to ρ but inversely proportional to A and l.
(b) directly proportional to ρ and l but inversely to A.
(c) directly proportional to ρ, A, and l.
(d) inversely proportional to ρ, A, and l.

The answer is (b)

Problem–15

The relationship between voltage, current, and resistance given by $V = IR$ is known as

- (a) Kirchhoff's current law.
- (b) Seebeck's voltage.
- (c) Ohm's law.
- (d) Thevenin's theorem.

The answer is (c)

Problem–16

The electromagnetic force obtained by joining the ends of wires of two dissimilar metals and keeping the joints at two different temperatures is known as

- (a) Kirchhoff's current law.
- (b) Seebeck's voltage.
- (c) Ohm's law.
- (d) Thevenin's theorem.

The answer is (b)

Problem–17

The fact that as much electrical current flows out of an electrical node as flows into it is called

- (a) Kirchhoff's current law.
- (b) Seebeck's voltage.
- (c) Ohm's law.
- (d) Thevenin's theorem.

The answer is (a)

Problem–18

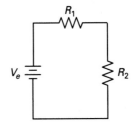

In the simple series (single loop) circuit shown in the accompanying illustration, which of the following statements is not true?

- (a) The current is the same through all circuit elements.
- (b) The equivalent resistance is the sum of the individual resistances.
- (c) The sum of the voltage drops across all components is equal to the equivalent applied voltage.
- (d) The voltage is the same through all the elements, e.g., $V_e = V_1 = V_2$.

The answer is (d)

Problem–19

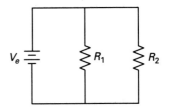

In the simple parallel circuit shown, which of the following statements is false?

- (a) The current is the same through all circuit elements.
- (b) The equivalent resistance is the sum of the individual resistances.
- (c) The sum of the voltage drops across all components is equal to the equivalent applied voltage.
- (d) The voltage is the same through all the elements, e.g., $V_e = V_1 = V_2$.

The answer is (d)

Problem–20

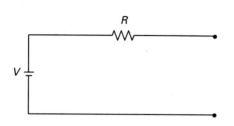

The circuit shown is known as which of the following?

  (a) Norton equivalent circuit
  (b) Thevenin equivalent circuit
  (c) voltage divider circuit
  (d) current divider circuit

The answer is (b)

Problem–21

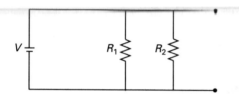

The circuit shown is known as which of the following?

  (a) Norton equivalent circuit
  (b) Thevenin equivalent circuit
  (c) simple parallel circuit
  (d) voltage divider circuit

The answer is (c)

Problem–22

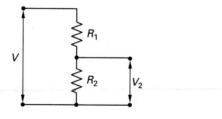

The circuit shown is known as which of the following?

  (a) Norton equivalent circuit
  (b) Thevenin equivalent circuit
  (c) simple parallel circuit
  (d) voltage divider circuit

The answer is (d)

Problem–23

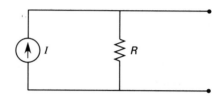

The circuit shown is known as which of the following?

  (a) Norton equivalent circuit
  (b) Thevenin equivalent circuit
  (c) simple parallel circuit
  (d) current divider circuit

The answer is (a)

Problem–24

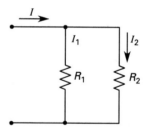

The circuit shown is known as which of the following?

  (a) Thevenin equivalent circuit
  (b) simple parallel circuit
  (c) voltage divider circuit
  (d) current divider circuit

The answer is (d)

Problem–25

The current present in any branch of a linear, single-voltage source network is interchangeable in location with the voltage source without affecting the current according to

    (a) the superposition theorem.
    (b) Norton's theorem.
    (c) the reciprocity theorem.
    (d) Thevenin's theorem.

The answer is (c)

Problem–26

The response of (i.e., voltage across or current through) a linear circuit element fed by two or more independent sources is equal to the response to each source taken individually with all other sources set to zero according to

    (a) the superposition theorem.
    (b) Norton's theorem.
    (c) the reciprocity theorem.
    (d) Thevenin's theorem.

The answer is (a)

Problem–27

A linear, two-terminal network with dependent or independent sources can be represented by an equivalent circuit consisting of a single-current source and resistor in parallel according to

    (a) the superposition theorem.
    (b) Norton's theorem.
    (c) the reciprocity theorem.
    (d) Thevenin's theorem.

The answer is (b)

Problem–28

A linear, two-terminal network with dependent and independent sources can be represented by a voltage source in series with a resistor according to

    (a) the superposition theorem.
    (b) Norton's theorem.
    (c) the reciprocity theorem.
    (d) Thevenin's theorem.

The answer is (d)

Problem–29

Any branch voltage can be replaced by a branch current, and any branch current can be replaced by a branch voltage, as long as the voltage drop and current through the branch remain the same according to

    (a) the superposition theorem.
    (b) Norton's theorem.
    (c) the reciprocity theorem.
    (d) the substitution theorem.

The answer is (d)

Problem–30

All of the following are recognized methods for determining current and voltage in electrical circuits except

    (a) Thevenin's theorem.
    (b) Norton's theorem.
    (c) the reciprocity theorem.
    (d) Kirchhoff's current and voltage laws.

The answer is (c)

Problem–31

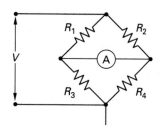

A Wheatstone bridge is commonly used to measure which of the following unknowns?

(a) resistance R
(b) voltage V
(c) current I
(d) power VI

The answer is (a)

Problem–32

In the transient behavior of RL and RC circuits, the time constant is the amount of time the circuit takes to reach what percentage of its steady-state value?

(a) 16.3%
(b) 24.3%
(c) 53.3%
(d) 63.3%

The answer is (d)

Problem–33

An Arsonval meter (galvanometer or permanent-magnet moving mechanism) is used to measure

(a) voltage.
(b) power.
(c) current.
(d) resistance.

The answer is (c)

Problem–34

The relationship between an AC current's frequency f and its angular frequency ω is given by

(a) $f = \dfrac{\omega}{\pi}$

(b) $\omega = 2\pi f$

(c) $f = 2\pi\omega$

(d) $\omega = \dfrac{f}{2\pi}$

The answer is (b)

Problem–35

All of the following terms are used to describe alternating waveforms except

(a) square.
(b) sawtooth.
(c) uniform or constant.
(d) sinusoidal.

The answer is (c)

Problem–36

The average value of an AC voltage symmetrical with the time axis is evaluated over half a cycle because

(a) this method is the accepted convention.
(b) the calculation is easier than over the full cycle.
(c) the average value over the full cycle is zero.
(d) the positive half of the cycle is the only contributor anyway.

The answer is (c)

Problem–37

What is the average value of the voltage of a rectified sinusoid expressed in terms of the mean value?

(a) zero

(b) $\dfrac{2V_m}{\pi}$

(c) $\dfrac{V_m}{\pi}$

(d) $\dfrac{V_m}{2}$

The answer is (b)

Problem–38

A DC current equal to the average value of a rectified AC current has the same electrolytic action for all of the following operations except

(a) capacitor charging.
(b) plating operations.
(c) heating.
(d) capacitor discharging.

The answer is (c)

Refer to the following illustration for Probs. 39 through 41.

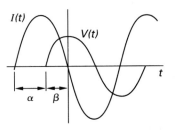

Problem–39

In the illustration shown, if an AC current $I(t)$ and voltage $V(t)$ are plotted as functions of time, the angle α is which of the following?

(a) wavelength
(b) phase angle difference
(c) phase angle
(d) angular frequency

The answer is (b)

Problem–40

In the illustration shown, if an AC current $I(t)$ and voltage $V(t)$ are plotted as functions of time, the angle β is which of the following?

(a) wavelength
(b) phase angle difference
(c) phase angle
(d) angular frequency

The answer is (c)

Problem–41

The current shown is which of the following?

(a) leading current
(b) lagging current
(c) resonant current
(d) DC current

The answer is (a)

Problem–42

A leading circuit has a phase angle difference that is

(a) positive.
(b) negative.
(c) zero.
(d) imaginary.

The answer is (a)

Problem–43

A circuit is called a leading circuit because

(a) voltage leads current.
(b) current leads voltage.
(c) current and voltage are in phase.
(d) current leads power.

The answer is (b)

Problem–44

A lagging circuit has a phase angle difference that is

(a) positive.
(b) negative.
(c) zero.
(d) imaginary.

The answer is (b)

Problem–45

A circuit is called a lagging circuit because

   (a) voltage leads current.
   (b) current leads voltage.
   (c) current and voltage are in phase.
   (d) voltage leads power.

The answer is (a)

Problem–46

The phase angle difference produced when a sinusoidal voltage is applied across an element alone (i.e., resistor, capacitor, or inductor) is called a(n)

   (a) phase angle.
   (b) impedance angle.
   (c) angular frequency.
   (d) inductance angle.

The answer is (b)

Use the following information for Probs. 47 through 49.

The following are characteristics of a typical electrical device:

  (I)   It has no resistance.
 (II)   It has no capacitance.
(III)   It has no inductance.
(IV)   It has a phase angle difference of $-90°$.
 (V)   It has a phase angle difference of $+90°$.

Problem–47

Which conditions are true for an ideal resistor?

   (a) I
   (b) I, II, and V
   (c) II and III
   (d) I, III, and IV

The answer is (c)

Problem–48

Which conditions are true for an ideal capacitor?

   (a) I
   (b) I, II, and V
   (c) II and III
   (d) I, III, and IV

The answer is (d)

Problem–49

Which conditions are true for an ideal inductor?

   (a) I
   (b) I, II, and V
   (c) II and III
   (d) I, III, and IV

The answer is (b)

Problem–50

The impedance equal to a combination of impedances in series is the

   (a) sum of the reciprocals of the impedances.
   (b) sum of the impedances.
   (c) difference of the impedances.
   (d) product of the impedances.

The answer is (b)

Problem–51

The impedance equal to a combination of impedances in parallel is the

   (a) inverse of the sum of the reciprocals of the impedances.
   (b) sum of the impedances.
   (c) difference of the impedances.
   (d) product of the impedances.

The answer is (a)

Problem–52

Which of the following statements about Ohm's law for AC circuits is not true?

(a) Only instantaneous values are considered.
(b) If the maximum values for the voltages are used, then maximum values for the current must be used.
(c) The arithmetic is complex.
(d) If the mean values for the voltages are used, then the mean values for the current must be used.

The answer is (a)

Problem–53

All of the following methods may be used with complex arithmetic for AC circuit analysis except

(a) Ohm's law.
(b) Kirchhoff's laws.
(c) node voltage methods.
(d) nonlinear superposition.

The answer is (d)

Problem–54

The reciprocal of the impedance Z is the complex quantity

(a) conductance, G.
(b) capacitance, C.
(c) admittance, Y.
(d) susceptance, B.

The answer is (c)

Problem–55

The reciprocal of the resistive part of impedance Z is

(a) conductance, G.
(b) capacitance, C.
(c) admittance, Y.
(d) susceptance, B.

The answer is (a)

Problem–56

The reciprocal of the reactive part of impedance Z is

(a) conductance, G.
(b) capacitance, C.
(c) admittance, Y.
(d) susceptance, B.

The answer is (d)

Problem–57

In a purely resistive circuit, the current

(a) leads the voltage by 90°.
(b) lags the voltage by 90°.
(c) is in phase with the voltage.
(d) leads the voltage by 45°.

The answer is (c)

Problem–58

In a purely capacitive circuit, the current

(a) leads the voltage by 90°.
(b) lags the voltage by 90°.
(c) is in phase with the voltage.
(d) can either lead or lag the voltage by 45°.

The answer is (a)

Problem–59

In a purely inductive circuit, the current

(a) leads the voltage by 90°.
(b) lags the voltage by 90°.
(c) is in phase with the voltage.
(d) can either lead or lag the voltage by 45°.

The answer is (b)

Problem–60

Which of the following statements about a circuit containing only an ideal capacitor is not true?

(a) The current leads the voltage.
(b) This is a leading circuit.
(c) The impedance is given by $Z = X_C \angle 90°$
(d) The voltage leads the current.

The answer is (d)

Problem–61

Which of the following statements about a circuit containing only an ideal inductor is not true?

(a) The current leads the voltage.
(b) This is a lagging circuit.
(c) The impedance is given by $Z = X_L \angle{-90°}$.
(d) The voltage leads the current.

The answer is (a)

Problem–62

Which of the following statements about a circuit containing only an ideal resistor is not true?

(a) The current is in phase with the voltage.
(b) This is an in phase circuit.
(c) The impedance angle ϕ is zero.
(d) The voltage leads the current.

The answer is (d)

Problem–63

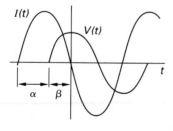

The illustration shows which of the following?

(a) a lagging phase angle difference
(b) a leading phase angle difference
(c) no phase angle difference
(d) a purely inductive circuit

The answer is (b)

Problem–64

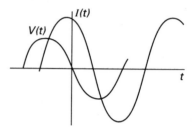

The illustration shows which of the following?

(a) a lagging phase angle difference
(b) a leading phase angle difference
(c) a purely resistive circuit
(d) a purely inductive circuit

The answer is (a)

Problem–65

For a series AC circuit, the total circuit impedance may be found as the

(a) sum of individual admittances.
(b) arithmetic sum of individual admittances.
(c) sum of the reciprocals of the individual impedances.
(d) sum of individual impedances.

The answer is (d)

Problem–66

For a parallel AC circuit consisting of inductances, capacitances, and/or resistances, the total circuit impedance may be found as the

(a) reciprocal of the sum of individual admittances.
(b) sum of the reciprocals of the individual impedances.
(c) arithmetic sum of individual admittances.
(d) sum of individual impedances.

The answer is (a)

Problem–67

For a series AC circuit, it is convenient to convert all known circuit element impedances to

(a) polar impedance form.
(b) exponential admittance form.
(c) rectangular impedance form.
(d) rectangular admittance form.

The answer is (c)

Problem–68

For a parallel AC circuit, it is convenient to convert all known circuit element impedances to

(a) polar impedance form.
(b) exponential admittance form.
(c) rectangular impedance form.
(d) rectangular admittance form.

The answer is (d)

Problem–69

For a series-RL circuit (no capacitance), the magnitude of the impedance is

(a) $\sqrt{R^2 + X_C^2}$
(b) $\sqrt{R^2 + (X_L - X_C)^2}$
(c) $\sqrt{R^2 + X_L^2}$
(d) $\sqrt{G^2 + B_L^2}$

The answer is (c)

Problem–70

For a parallel RC circuit (no inductance), the magnitude of the admittance can be found as

(a) $\sqrt{R^2 + X_C^2}$
(b) $\sqrt{R^2 + (X_L - X_C)^2}$
(c) $\sqrt{R^2 + X_L^2}$
(d) $\sqrt{G^2 + B_C^2}$

The answer is (d)

Problem–71

A resonant circuit has a

(a) positive phase angle difference.
(b) negative phase angle difference.
(c) zero phase angle difference.
(d) positive impedance angle.

The answer is (c)

Problem–72

Resonance in a circuit can be obtained by

(a) adjusting the frequency for a circuit of fixed elements.
(b) either adjusting the frequency or adjusting the capacitive and inductive elements in the circuit.
(c) removing resistances in the circuit.
(d) adding resistances to the circuit.

The answer is (b)

Problem–73

In a resonant series-RLC circuit, all of the following are true except

(a) impedance equals resistance.
(b) current and voltage are in phase.
(c) current is a minimum.
(d) power dissipation is a maximum.

The answer is (c)

Problem–74

In a resonant parallel *RLC* circuit, all of the following are true except

  (a) impedance equals resistance.
  (b) current and voltage are in phase.
  (c) current is a minimum.
  (d) power dissipation is a maximum.

The answer is (d)

Problem–75

A resonant circuit has a

  (a) large and positive phase angle difference.
  (b) small and positive phase angle difference.
  (c) zero phase angle difference.
  (d) small and negative phase angle difference.

The answer is (c)

Problem–76

At resonance the relationship between reactive and capacitive impedance is most correctly

  (a) $X_L = X_C + R$.
  (b) $X_L < X_C$.
  (c) $X_L = X_C$.
  (d) $X_L > X_C$.

The answer is (c)

Problem–77

The frequency at which an AC circuit becomes purely resistive is called the

  (a) nonresonant frequency.
  (b) capacitive frequency.
  (c) inductive frequency.
  (d) resonant frequency.

The answer is (d)

Problem–78

At frequencies below the resonant frequency, a series-*RLC* circuit is

  (a) resonant.
  (b) capacitive.
  (c) inductive.
  (d) resistive.

The answer is (b)

Problem–79

At frequencies above the resonant frequency, a series-*RLC* circuit is

  (a) resonant.
  (b) capacitive.
  (c) inductive.
  (d) resistive.

The answer is (c)

Problem–80

The only real energy dissipation in an AC circuit occurs in

  (a) capacitors.
  (b) resistors.
  (c) inductors.
  (d) capacitors and inductors.

The answer is (b)

Problem–81

The power factor is expressed in terms of the power angle ϕ (also known as the impedance angle ϕ) as

  (a) ϕ.
  (b) $\sin \phi$.
  (c) $\cos \phi$.
  (d) $\tan \phi$.

The answer is (c)

Problem–82

For a purely resistive load, what is the value of the power angle ϕ?

(a) $\dfrac{\pi}{2}$

(b) $\dfrac{\pi}{4}$

(c) 0

(d) $-\dfrac{\pi}{2}$

The answer is (c)

Problem–83

For a purely reactive load, what is the value of the power factor?

(a) 0

(b) 0.31

(c) 0.81

(d) 1.00

The answer is (a)

Problem–84

How can the magnitude of the complex power S be expressed, in terms of real power P and reactive power Q?

(a) $S^2 = Q^2 - P^2$

(b) $S^2 = P^2 - Q^2$

(c) $S^2 = P^2 + Q^2$

(d) $S = P + Q$

The answer is (c)

Problem–85

If the turns ratio of a transformer (the ratio of numbers of primary to secondary windings) is greater than unity, the transformer

(a) increases voltage.

(b) decreases voltage.

(c) maintains the same voltage.

(d) can either increase or decrease the voltage.

The answer is (b)

Problem–86

In general, in transient analyses of RC and RL circuits, steady state has been reached after how many time constants?

(a) 1

(b) 2

(c) 3

(d) 5

The answer is (d)

Problem–87

The amount of voltage induced when some of the magnetic flux produced in one coil passes through a second coil is given by

(a) Kirchhoff's current law.

(b) Kirchhoff's voltage law.

(c) Ohm's law.

(d) Faraday's law.

The answer is (d)

Problem–88

Silicon (Si) and germanium (Ge), which form the basis of most semiconductors, are how much more or less conductive than insulators?

(a) much more

(b) slightly more

(c) the same

(d) much less

The answer is (b)

Problem–89

An increase in conductivity of semiconductors can be affected by the addition of minute amounts (e.g., 10 ppb) of

(a) pure metal.

(b) inert gas.

(c) dopes.

(d) sodium chloride.

The answer is (c)

Problem–90

All of the following doping elements have three valence electrons except

  (a) aluminum.
  (b) boron.
  (c) gallium.
  (d) phosphorus.

The answer is (d)

Problem–91

The intrinsic behavior of a semiconductor is caused by the existence of

  (a) ions.
  (b) isomorphs.
  (c) crystalline defects.
  (d) isotopes.

The answer is (c)

Problem–92

Semiconductor devices are inherently

  (a) linear but analyzed as nonlinear.
  (b) nonlinear but analyzed as linear.
  (c) linear and analyzed as linear.
  (d) constant voltage devices.

The answer is (b)

Problem–93

Semiconductors manufactured with acceptor dopes to give positive holes as majority carriers are called

  (a) p-types.
  (b) n-types.
  (c) np-types.
  (d) pnp-types.

The answer is (a)

Problem–94

Semiconductors manufactured with acceptor dopes to give electrons as majority carriers are called

  (a) p-types.
  (b) n-types.
  (c) pn-types.
  (d) pnp-types.

The answer is (b)

Problem–95

A device which passes current in only one direction is a(n)

  (a) ideal diode.
  (b) resistor.
  (c) capacitor.
  (d) inductor.

The answer is (a)

Problem–96

All of the following are typical diode applications except a

  (a) half-wave rectifier.
  (b) full-wave rectifier.
  (c) peak clipper.
  (d) integrator.

The answer is (d)

Problem–97

A bipolar junction transistor (BJT) consists of

  (a) a thin n-type semiconductor between two p-type semiconductors.
  (b) a thin p-type semiconductor between two n-type semiconductors.
  (c) a thin n-type semiconductor between two p-type semiconductors, or a thin p-type semiconductor between two n-type semiconductors.
  (d) two n-type semiconductors (no p-type).

The answer is (c)

Refer to the following illustrations for Probs. 98 through 102.

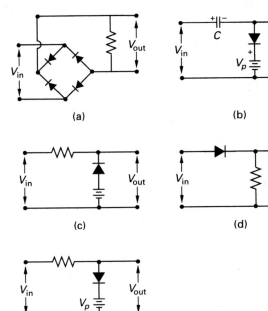

(a)   (b)

(c)   (d)

(e)

Problem–98

Part (a) illustrates a

   (a) half-wave rectifier.
   (b) full-wave rectifier.
   (c) clamping circuit.
   (d) base clipper.

The answer is (b)

Problem–99

Part (b) illustrates a

   (a) half-wave rectifier.
   (b) full-wave rectifier.
   (c) clamping circuit.
   (d) base clipper.

The answer is (c)

Problem–100

Part (c) illustrates a

   (a) half-wave rectifier.
   (b) full-wave rectifier.
   (c) base clipper.
   (d) peak clipper.

The answer is (c)

Problem–101

Part (d) illustrates a

   (a) half-wave rectifier.
   (b) full-wave rectifier.
   (c) base clipper.
   (d) peak clipper.

The answer is (a)

Problem–102

Part (e) illustrates a

   (a) half-wave rectifier.
   (b) clamping circuit.
   (c) base clipper.
   (d) peak clipper.

The answer is (d)

Refer to the following illustrations for Probs. 103 through 107.

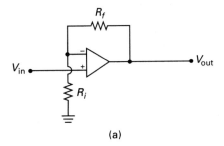

(a)

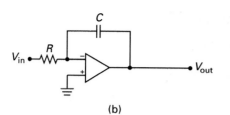

(b)

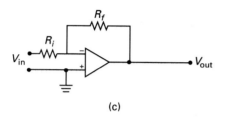

(c)

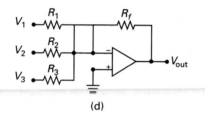

(d)

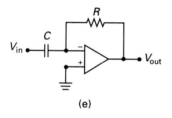

(e)

Problem–103

Part (a) illustrates what kind of operational amplifier application?

(a) inverting amplifier
(b) noninverting amplifier
(c) integrator
(d) differentiator

The answer is (b)

Problem–104

Part (b) illustrates what kind of operational amplifier application?

(a) inverting amplifier
(b) summing amplifier
(c) integrator
(d) differentiator

The answer is (c)

Problem–105

Part (c) illustrates what kind of operational amplifier application?

(a) inverting amplifier
(b) noninverting amplifier
(c) integrator
(d) differentiator

The answer is (a)

Problem–106

Part (d) illustrates what kind of operational amplifier application?

(a) inverting amplifier
(b) noninverting amplifier
(c) summing amplifier
(d) integrator

The answer is (c)

Problem–107

Part (e) illustrates what kind of operational amplifier application?

(a) inverting amplifier
(b) summing amplifier
(c) integrator
(d) differentiator

The answer is (d)

Problem–108

An inductor

   (a) stores electric charge.
   (b) impedes a change in current flow.
   (c) impedes current flow.
   (d) increases current flow.

The answer is (b)

Problem–109

A capacitor

   (a) stores electric charge.
   (b) impedes a change in current flow.
   (c) impedes current flow.
   (d) increases current flow.

The answer is (a)

Problem–110

A resistor

   (a) stores electric charge.
   (b) impedes a change in current flow.
   (c) impedes current flow.
   (d) increases current flow.

The answer is (c)

Chapter 10
Engineering Economics

Problem–1

Which of the following is not an assumption implicit in solving engineering economic analysis problems?

  (a) The year-end convention is applicable.

  (b) There is no inflation now, nor will there be any during the lifetime of the project.

  (c) The time value of money is zero.

  (d) Unless specifically stated otherwise, a before-tax analysis is needed.

The answer is (c)

Problem–2

Let F be future worth, A be a uniform annual amount, i be the effective interest per compounding period, and n be the number of compounding periods. Which of the following correctly relates these variables?

  (a) $F = A\left[\dfrac{(1+i)^n - 1}{i(1+i)^n}\right]$

  (b) $F = A\left[\dfrac{(1+i)^n - 1}{i}\right]$

  (c) $A = F(1+i)^n$

  (d) $F = A(1+i)^{-n}$

The answer is (b)

Problem–3

Let P be present worth, A be a uniform annual amount, i be the effective interest per compounding period, and n be the number of compounding periods. Which of the following correctly relates these variables?

  (a) $A = P(1+i)^n$

  (b) $P = A\left[\dfrac{(1+i)^n - 1}{i(1+i)^n}\right]$

  (c) $P = A(1+i)^{-n}$

  (d) $A = P\left[\dfrac{(1+i)^n - 1}{i}\right]$

The answer is (b)

Problem–4

The present worth of an infinite (perpetual) series of annual amounts is given by which of the following?

  (a) $P = A\left[\dfrac{(1+i)^n - 1}{i(1+i)^n}\right]$

  (b) $P = \dfrac{A}{i}$

  (c) $P = A(1+i)^{-n}$

  (d) $P = A(1+i)^n$

The answer is (b)

Problem–5

The rule of 72 can be used to calculate the

  (a) number of monthly payments in six years.

  (b) time to double an investment.

  (c) time to triple an investment.

  (d) time to decrease an investment to $1/72$ of its original value.

The answer is (b)

Problem–6

A sinking fund is a(n)

  (a) annual amount equal to all cash flows.

  (b) fund or account into which annual deposits of A are made to accumulate F at the time $t = n$ in the future.

  (c) series of payments made over a period of time.

  (d) equivalent uniform annual cost.

The answer is (b)

Problem–7

All of the following are varied and nonstandard cash flows except

(a) gradient cash flow.
(b) stepped cash flow.
(c) delayed or premature cash flow.
(d) uniform (standard) cash flow.

The answer is (d)

Problem–8

One item that differentiates the "Christmas club problem" from a standard cash flow problem is

(a) varying interest rates through the life of the problem.
(b) varying cash flows through the life of the problem.
(c) the existence of a payment at the beginning of the time at $t = 0$ and the absence of a payment at the end of the problem.
(d) the nonexistence of interest in the problem at all.

The answer is (c)

Problem–9

Compound interest differs from simple interest because

(a) compound interest is not taxable in Nevada.
(b) simple interest cannot be applied to multiple bank accounts.
(c) the interest earns interest.
(d) compound interest accounts are insured by an agency of the federal government.

The answer is (c)

Problem–10

Rate of return (ROR) is

(a) a dollar amount.
(b) the minimum attractive rate of return.
(c) an effective annual interest rate.
(d) synonymous to return on investment (ROI).

The answer is (c)

Problem–11

Return on investment (ROI) is

(a) a dollar amount.
(b) the minimum attractive rate of return.
(c) an effective annual interest rate.
(d) synonymous to rate of return (ROR).

The answer is (a)

Problem–12

One of the advantages of comparing interest rates to the minimum attractive rate of return (MARR) is that

(a) the MARR is easy to calculate.
(b) an effective interest rate need never be known.
(c) it need not be used in numerical calculations.
(d) it is guaranteed by an agency of the federal government.

The answer is (b)

Problem–13

Typical alternative comparison problem formats are characterized by all except one of the following.

(a) An investment rate will be given.
(b) Two or more alternatives will be competing for funding.
(c) Each alternative will have its own cash flows.
(d) All interest rates are expressed as whole numbers.

The answer is (d)

Problem–14

All of the following are possible methods for selecting the best alternative from a group of proposals except

(a) future cost method.
(b) break-even analysis.
(c) the present worth method.
(d) the annual cost method.

The answer is (b)

Problem–15

The equivalent uniform annual cost (EUAC) is very helpful in

(a) calculating retirement plan incomes.
(b) comparing alternatives with different lives.
(c) computing future values of Christmas club accounts.
(d) determining the present value of an annuity.

The answer is (b)

Problem–16

An opportunity cost is a(n)

(a) salvage cost.
(b) imaginary cost representing what will not be received if a particular strategy is rejected.
(c) trade-in allowance.
(d) equivalent annual cost for charity contributions.

The answer is (b)

Problem–17

An artificial expense that spreads the purchase price of an asset or other property over a number of years is called

(a) life-cycle costing.
(b) depreciation.
(c) amortization.
(d) depletion.

The answer is (b)

Problem–18

An artificial deductible operating expense designed to compensate mining organizations for decreasing mineral reserves is called

(a) life-cycle costing.
(b) depreciation.
(c) amortization.
(d) depletion.

The answer is (d)

Problem–19

An artificial expense that spreads the cost of an asset over some base (time, units of production, number of customers, etc.) is

(a) life-cycle costing.
(b) amortization.
(c) depletion.
(d) book value.

The answer is (b)

Problem–20

The difference between the original purchase price and accumulated depreciation is the

(a) life-cycle costing.
(b) amortization.
(c) depletion.
(d) book value.

The answer is (d)

Problem–21

All of the following are depreciation methods except

(a) straight line.
(b) constant percentage.
(c) double declining balance.
(d) capitalizing the asset.

The answer is (d)

Refer to the following illustration for Probs. 22 and 23.

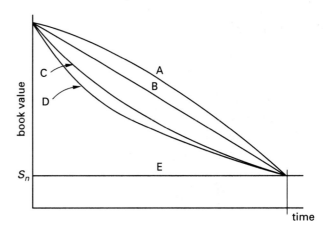

Problem–22

The accelerated depreciation method(s) is (are) given by which curve(s)?

   (a) A
   (b) C
   (c) C and D
   (d) A and D

The answer is (c)

Problem–23

The sinking fund depreciation method(s) is (are) given by which curve(s)?

   (a) A
   (b) C
   (c) C and D
   (d) A and D

The answer is (a)

Problem–24

The sinking fund method is seldom used in industry because

   (a) the term "sinking fund" has a bad connotation for business.
   (b) sometimes it gives misleading answers.
   (c) the initial depreciation is low.
   (d) the initial depreciation is high.

The answer is (c)

Problem–25

A depreciation method that does not consider the salvage value of an asset is known as

   (a) amortization.
   (b) an accelerated depreciation method.
   (c) an unadjusted basis.
   (d) an accumulated depreciation method.

The answer is (c)

Problem–26

Depreciation is used because

   (a) it is fair for small businesses.
   (b) tax regulations do not allow the cost of an asset as a deductible expense in the year of purchase.
   (c) it is easier to compute than a single expense.
   (d) it is fair for large businesses.

The answer is (b)

Problem–27

A common depreciation basis is the

   (a) purchase price of an asset.
   (b) salvage value of an asset.
   (c) difference between the purchase price and salvage value of an asset.
   (d) difference between purchase price and book value.

The answer is (c)

Problem–28

If k is the number of compounding periods per year, and ϕ is the effective rate per compounding period, then how can the effective annual rate i be calculated?

   (a) $(1 + \phi)^k$
   (b) $(1 + \phi)^k - 1$
   (c) $(1 + \phi)^{-k}$
   (d) $(1 + \phi)^{-k} + 1$

The answer is (b)

Problem–29

Let F be the future worth, P the present value, and r the nominal interest rate per year. The present value can be found from the future value for the case of continuous compounding by multiplying by which factor?

   (a) $(1 + r)^n$
   (b) $(1 + 4)^{-n}$
   (c) e^{-rn}
   (d) e^{rn}

The answer is (c)

Problem–30

A cost that is a function of an independent variable (e.g., number of units) is called a(n)

(a) fixed cost.
(b) incremental cost.
(c) variable cost.
(d) administrative expense.

The answer is (c)

Refer to the following illustration for Probs. 31 and 32.

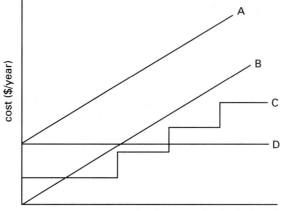

Problem–31

Variable costs are shown by which curve?

(a) A
(b) B
(c) C
(d) D

The answer is (b)

Problem–32

The sum of fixed and variable costs is shown by which curve?

(a) A
(b) B
(c) C
(d) D

The answer is (a)

Problem–33

All of the following are usually considered fixed costs except

(a) rent.
(b) interest on loans.
(c) direct labor costs.
(d) property taxes.

The answer is (c)

Problem–34

All of the following are usually considered variable costs except

(a) cost of miscellaneous supplies.
(b) payroll benefit costs.
(c) income taxes.
(d) janitorial service expenses.

The answer is (d)

Problem–35

A method of telling when the economic value of one alternative becomes equal to the economic value of another alternative is called

(a) cost-benefit analysis.
(b) cost accounting.
(c) break-even analysis.
(d) accelerated depreciation.

The answer is (c)

Problem–36

The book value of an asset equals its salvage value at the end of the asset's depreciation period in all of the following depreciation methods except

- (a) sinking fund.
- (b) straight line.
- (c) double declining balance.
- (d) accelerated cost recovery system (ACRS).

The answer is (c)

Problem–37

An accelerated depreciation method is one which

- (a) calculates the depreciation amount faster than other methods when automated for a computer.
- (b) calculates a depreciation amount greater than the straight line method.
- (c) calculates a depreciation amount less than the straight line method.
- (d) calculates a greater market value of an asset than does the straight line method.

The answer is (b)

Problem–38

The payback period is defined as the

- (a) length of time, usually in years, for the cumulative net annual income to equal the initial investment.
- (b) length of time, usually in years, for the cumulative annual decrease in expenses due to the purchase of the asset to equal the initial investment.
- (c) length of time to pay back the loan for the asset.
- (d) length of time, usually in years, for the cumulative net annual profit to equal the initial investment.

The answer is (d)

Problem–39

Almost all engineering economics analyses avoid the need to consider the effects of inflation or a changing price index. However, if the effective annual interest rate i must be corrected for the annual inflation rate e for n years, how is the corrected value i' calculated?

- (a) $(1+i)^n$
- (b) $i + e + ie$
- (c) $(1+1)^{-n}$
- (d) $(1+e)^n$

The answer is (b)

Problem–40

Almost all engineering economics analyses avoid the need to consider the effects of inflation or a changing price index. However, if all cash flows must be converted to constant value dollars, the effective annual interest rate i, the annual inflation rate e, and the number of years n may be combined so as to divide by which of the following?

- (a) $(1+i)^n$
- (b) $i + e + ie$
- (c) $(1+1)^{-n}$
- (d) $(1+e)^n$

The answer is (d)

Problem–41

The economic order quantity is the

- (a) minimum order size necessary for bulk discounts.
- (b) order quantity that minimizes inventory costs per unit time.
- (c) order quantity that minimizes fixed costs per unit time.
- (d) order quantity that minimizes the depletion rate per unit time.

The answer is (b)

Problem–42

All of the following are nonquantifiable factors of engineering economic analyses except

    (a) preferences.
    (b) political ramifications.
    (c) salvage value.
    (d) goodwill.

The answer is (c)

Problem–43

All of the following are standard cash flows except

    (a) single payment.
    (b) uniform series.
    (c) derivative.
    (d) exponential gradient.

The answer is (c)

Problem–44

One of the simplest, most effective methods for handling cash flows with missing parts or extra parts (compared to a standard cash flow) is

    (a) integration.
    (b) differentiation.
    (c) superposition.
    (d) partial fractions.

The answer is (c)

Problem–45

One simple mnemonic device for remembering that $F = P \times F/P$ (i.e., the future value equals the present value times the discounting factor of the future value for a given present value) is to think of the discounting factors (e.g., F/P) as

    (a) proper fractions.
    (b) conditional probabilities.
    (c) partial fractions.
    (d) complex fractions.

The answer is (b)

Problem–46

Economic engineering analyses can be conducted accurately in

    (a) British pounds.
    (b) American dollars.
    (c) Japanese yen.
    (d) any monetary unit.

The answer is (d)

Problem–47

Which of the following is not an assumption implicit in engineering economic analysis?

    (a) The beginning of the year convention is applicable.
    (b) There is no inflation now, nor will there be any inflation during the lifetime of the project.
    (c) The effective interest rate in the problem will be constant during the lifetime of the project.
    (d) Nonquantifiable factors can be disregarded.

The answer is (a)

Problem–48

The tabulated functions that relate the present worth P, the future worth F, the annual amount A, the effective interest rate i, and the number of compounding periods n, (e.g., $(F/P, i\%, n) = (1 + i)^n$) are known as

    (a) proper fractions.
    (b) improper fractions.
    (c) discounting factors.
    (d) annuities.

The answer is (c)

Problem–49

Economic conclusions based upon engineering economic analyses can be reached

(a) by one and only one procedure.
(b) by more than one method.
(c) only by those licensed by the state.
(d) only during the current fiscal year.

The answer is (b)

Problem–50

A method for helping to visualize and simplify problems having diverse receipts and disbursements is

(a) Lotus spreadsheet analysis.
(b) Excel spreadsheet analysis.
(c) cash flow diagram.
(d) having an accountant in the family.

The answer is (c)

Problem–51

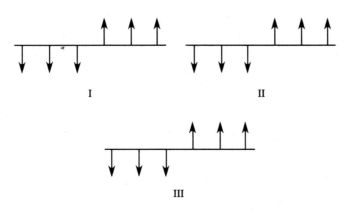

The cash flow with the greatest present value is which of the following?

(a) I
(b) II
(c) III
(d) It cannot be determined without knowing the interest rates.

The answer is (d)

Problem–52

The basic accounting equation is

(a) assets + liability = owners' equity.
(b) assets = liability + owners' equity.
(c) assets + owners' liability = equity.
(d) assets = liability.

The answer is (b)

Problem–53

All of the following are accepted methods for determining the cost elements in an inventory except

(a) first-in, last-out (FILO).
(b) average cost.
(c) first-in, first-out (FIFO).
(d) last-in, first-out (LIFO).

The answer is (a)

Problem–54

Engineering economic analysis that addresses variables with a known or estimated probability distribution is called

(a) synthetical analysis.
(b) risk analysis.
(c) sensitivity analysis.
(d) conditional probability analysis.

The answer is (b)

Problem–55

Engineering economic analysis that is concerned with situations in which there is not enough information to determine probability or frequency for the variables involved is

(a) synthetical analysis.
(b) risk analysis.
(c) sensitivity analysis.
(d) uncertainty analysis.

The answer is (d)

Chapter 11
Ethics

Problem—1

One of the members of an engineering firm suggests that the firm send Christmas gifts to some of their business associates. The company president rejects the idea because it would violate ethical codes in all but one case. Of the following, who could the company send gifts to?

(a) The building contractor it worked with on its last major job.

(b) A vendor who provides many of its supplies.

(c) A company it is trying to attract as a client.

(d) The chairman of the state board that awards most of its work.

The answer is (a)

Problem—2

Which of the following is not a reason for the existence of ethical codes?

(a) Engineers possess special knowledge and training, which makes it easy for them to take advantage of clients and the public.

(b) The services performed by engineers are vitally important to the public welfare.

(c) Without guidelines, most people behave unethically.

(d) Guidelines help resolve ethical problems.

The answer is (c)

Problem—3

It is ethical for other engineers to review your work only if

(a) they are fully competent in the field.

(b) they sign a consulting contract with you that includes a nondisclosure clause.

(c) they are a licensed professional engineer.

(d) your client knows and approves of their participation.

The answer is (d)

Problem—4

The primary reason for the existence of ethical codes is so that

(a) engineers can tell right from wrong.

(b) the government can prosecute ethical transgressions.

(c) minimum standards of acceptable practice can be established.

(d) clients can evaluate an engineer's performance.

The answer is (c)

Problem—5

Ethical requirements are satisfied most completely by

(a) doing what makes you feel good.

(b) observing all applicable laws.

(c) following the instructions of your client/employer.

(d) doing what will bring the most good to the most people.

The answer is (d)

Problem-6

Which of the following is not prohibited by ethical codes?

(a) coordinating a project that includes design segments for which the engineer is unqualified
(b) contributing to the campaign of a politician who is likely to be in a position to award projects to the engineer
(c) accepting compensation for a project from more than one source without making full disclosure to all parties
(d) approving a design that has the possibility of endangering the public

The answer is (a)

Problem-7

Which of the following is not required of engineers by ethical codes?

(a) informing proper authorities of possible ethical violations
(b) avoiding business associations with public officials
(c) maintaining confidentiality of proprietary information
(d) identifying any party on whose behalf they issue a public statement

The answer is (a)

Problem-8

Which of the following does not constitute a conflict of interest?

(a) An engineer whose firm is competing for a government contract is asked to participate on the advisory board that will award the contract.
(b) A professional athlete bets on games in which he will play.
(c) A politician with significant investment in a certain construction company is chosen to head the committee awarding highway contracts.
(d) An employee of an architecture firm moonlights as a computer programmer, using the company's workstations to develop graphical routines.

The answer is (d)

Problem-9

Which of the following is not required for an engineer to accept a project?

(a) The engineer must be fully competent in the area with which the project is concerned.
(b) The project must have been won through a competitive bidding process.
(c) The project must in no way endanger the public.
(d) The client must be informed of any possible conflicts of interest that the engineer might have.

The answer is (b)

Problem-10

Which of the following is not a goal of ethical codes?

(a) to protect the public welfare
(b) to ensure competence among registered engineers
(c) to promote uniformity of services offered by engineers
(d) to prevent unfair business practices

The answer is (c)

Problem-11

Which of the following factors is least likely to affect the application and interpretation of ethical codes in a country?

(a) the country's political system
(b) the country's economic system
(c) the religious environment
(d) the country's language

The answer is (d)

Problem-12

Which of the following projects would it probably be unethical for a civil engineer to design?

(a) a wastewater treatment plant
(b) the elevator system for a high-rise office building
(c) an airport's runway layout
(d) the seismic retrofit of a suspension bridge

The answer is (b)

Problem–13

All of the following are considered to be professions except

(a) engineering.
(b) law.
(c) medicine.
(d) politics.

The answer is (d)

Problem–14

Which of the following is not a characteristic of the learned professions?

(a) They require extensive training.
(b) They make extensive use of members' independent judgment.
(c) They are regulated by ethical standards.
(d) They are answerable only to their regulatory agencies.

The answer is (d)

Problem–15

What body (bodies) is (are) responsible for the registration of professional engineers?

(a) the federal government
(b) state governments
(c) NCEES
(d) professional engineering societies

The answer is (b)

Problem–16

The application of ethical codes is

(a) subject to interpretation.
(b) clearly defined.
(c) uniform from state to state.
(d) independent of the circumstances.

The answer is (a)

Problem–17

Engineers may reveal proprietary information

(a) after their period of employment is completed.
(b) after a period of time specified in their contract.
(c) as long as their new employer is not a competitor of their former employer.
(d) if other ethical considerations demand it.

The answer is (d)

Problem–18

Which of the following is not necessarily required by ethics for an engineer to make a public statement? The engineer must

(a) be knowledgeable of the subject.
(b) reveal his/her interest in the matter.
(c) be objective and truthful.
(d) have the permission of his/her client/employer.

The answer is (d)

Problem–19

Which of the following principles is not embodied in ethical codes?

(a) Engineers' foremost responsibility is protection of the public welfare.
(b) Engineers should provide services only in their specific areas of technical competence.
(c) Engineers must maintain membership in professional societies to ensure that they maintain their level of competence.
(d) Engineers should avoid any conflicts of interest.

The answer is (c)

Problem–20

Which of the following situations would not be perceived as a conflict of interest?

- (a) An engineer working for a firearms manufacturer holds an advisory position with the National Rifle Association.
- (b) A local contractor offers to pay all living expenses for a field engineer while he is overseeing a project.
- (c) Several of the members of a committee investigating a plane crash are on the board of directors of the plane's manufacturer.
- (d) A technical expert testifying before a commission investigating the feasibility of an airport expansion is offered a consulting job by one of the contractors, scheduled to begin the day after the commission reports its findings.

The answer is (a)

Chapter 12
Computer Engineering

Problem–1

How would you express the address of the cell located at the intersection of row 28 and column 27?

  (a) AA28
  (b) AB28
  (c) 28.27
  (d) 27.AB

The answer is (a)

Problem–2

In a typical spreadsheet, the reference H$8 is typed into cell T5. If this reference is copied into cell AA8, which cell will it refer to?

  (a) AA5
  (b) O8
  (c) O11
  (d) T8

The answer is (b)

Problem–3

Flowcharts usually begin with which symbol?

  (a) annotation
  (b) input/output
  (c) process
  (d) terminal

The answer is (d)

Problem–4

What is the value of the following logical expression when the boolean variables A, B, and C are set to TRUE, TRUE, and FALSE, respectively?

$$(.NOT.(A.OR.B)).AND.(B.AND.C)$$

  (a) TRUE
  (b) FALSE
  (c) neither TRUE nor FALSE
  (d) both TRUE and FALSE

The answer is (b)

Problem–5

Which control structure is used to execute a block of code a specified number of times?

  (a) DO...WHILE
  (b) FOR...NEXT
  (c) GOTO
  (d) IF...THEN

The answer is (b)

Problem–6

Which control structure is used to decide between two alternate courses of action?

  (a) DO...WHILE
  (b) FOR...NEXT
  (c) GOTO
  (d) IF...THEN

The answer is (d)

Problem-7

Which of the following control structures are branching statements?

(I)   FOR...NEXT

(II)   GOTO

(III)   IF...THEN

  (a) I only
  (b) II only
  (c) I and II
  (d) II and III

The answer is (b)

Problem-8

How many bits are there in a megabyte?

  (a) 2^{12}
  (b) 2^{18}
  (c) 2^{20}
  (d) 2^{23}

The answer is (d)

Problem-9

What is the shortest time that a 28,800 bps modem can transmit a 1 MB file?

  (a) 36 s
  (b) 290 s
  (c) 980 s
  (d) 2300 s

The answer is (b)

Problem-10

What is the square of the binary number 1010 (in binary form)?

  (a) 100
  (b) 101000
  (c) 1100100
  (d) 10101010

The answer is (c)

Problem-11

A modem

  (a) converts data from digital to analog signals and back again.
  (b) communicates between CPU and peripherals.
  (c) manages data transmission between CPU and storage devices.
  (d) stores data which must be accessed frequently.

The answer is (a)

Problem-12

Computational mathematics is based on powers of which of the following?

  (a) 2
  (b) 8
  (c) 10
  (d) 12

The answer is (a)

Problem-13

What component of a typical computer system has the largest memory capacity?

  (a) RAM
  (b) ROM
  (c) hard disk
  (d) floppy disk

The answer is (c)

Problem-14

The smallest piece of information that can be interpreted by a computer is the

  (a) bit.
  (b) byte.
  (c) character.
  (d) word.

The answer is (a)

Problem—15

If the formula A4 + B4 is typed into cell C4 and then copied into cell E5, what sum would be represented in cell E5?

    (a) A4 + B4
    (b) A5 + B5
    (c) B4 + C5
    (d) C5 + D5

The answer is (d)

Problem—16

Which of the following control structures are looping statements?

  (I)   DO...WHILE

 (II) FOR...NEXT

(III) IF...THEN

    (a) I only
    (b) III only
    (c) I and II
    (d) II and III

The answer is (c)

Problem—17

If A = TRUE and B = FALSE, which of the following expressions evaluates to TRUE?

    (a) A.AND.B
    (b) .NOT.(A.OR.B)
    (c) A.OR.(A.AND.B)
    (d) B.AND.(A.OR.B)

The answer is (c)

Problem—18

A pseudocode program contains the following block. If the final value of RESULT is TRUE, what is the value of the boolean variable A?

```
B = TRUE
C = FALSE
IF A.AND.C THEN B = .NOT.B
IF A.AND.B THEN RESULT = TRUE
ELSE RESULT = FALSE
```

    (a) TRUE
    (b) FALSE
    (c) either TRUE or FALSE
    (d) neither TRUE or FALSE

The answer is (c)

Problem—19

Which of the following is a volatile data storage device?

    (a) hard disk
    (b) CD-ROM
    (c) WORM drive
    (d) RAM

The answer is (d)

Problem—20

What is the address of the cell located directly to the left of cell K6 in a typical spreadsheet program?

    (a) J5
    (b) J6
    (c) K5
    (d) L7

The answer is (b)

More FE/EIT Exam Practice!

EIT Review Manual
Rapid Preparation for the General Fundamentals of Engineering Exam

Michael R. Lindeburg, PE.
Paperback

The *EIT Review Manual* prepares you for the FE/EIT exam in the fastest, most efficient way possible. Updated to reflect the new exam content and format, this book gives you diagnostic tests to see what you need to study most, concise reviews of all exam topics, 1000+ practice problems with solutions, a complete eight-hour sample exam, and FE/EIT quiz software on diskette for your PC. The *EIT Review Manual* is your best choice if you are still in college or a recent graduate, or if your study time is limited.

Engineer-In-Training Reference Manual

Michael R. Lindeburg, PE.
Hardcover

The "Big Yellow Book" has been helping engineers pass the FE/EIT exam for more than 20 years. A tried-and-true review, it provides in-depth coverage of subjects typically found in undergraduate engineering programs, focusing especially on those included on the FE/EIT exam. Thousands of formulas, tables, and illustrations plus more than 900 practice problems make the *Engineer-In-Training Reference Manual* the most comprehensive study manual you can buy—and an ideal choice if you have been out of school for a while.

Engineer-In-Training Sample Examinations

Michael R. Lindeburg, PE.
Paperback

Most examinees benefit by taking a practice run through the FE/EIT exam before sitting down for the real thing. This book contains two full-length practice exams, including detailed solutions to all the problems. Each of these eight-hour tests has typical morning and general afternoon questions in multiple-choice format. Working these sample exams is a great way to get extra practice for the FE/EIT exam.

Calculus Refresher for the Fundamentals of Engineering Exam

Peter Schiavone, PhD.
Paperback

Many engineers report having more trouble with problems involving calculus than with anything else on the FE/EIT exam. This book covers all the areas that you'll need to know for the exam: differential and integral calculus, centroids and moments of inertia, differential equations, and precalculus topics such as quadratic equations and trigonometry. You get clear explanations of theory, relevant examples, and FE-style practice problems (with solutions). If you are at all unsure of your calculus skills, this book is a must.

To order: 1-800-426-1178 or www.ppi2pass.com